SCIENCE WITHOUT LAWS

SCIENCE AND CULTURAL THEORY

A Series Edited by Barbara Herrnstein Smith
and E. Roy Weintraub

SCIENCE WITHOUT LAWS

MODEL SYSTEMS, CASES, EXEMPLARY NARRATIVES

Edited by Angela N. H. Creager, Elizabeth Lunbeck, and M. Norton Wise

Duke University Press Durham and London 2007

© 2007 Duke University Press
All rights reserved
Printed in the United States of
America on acid-free paper ∞
Designed by Amy Ruth Buchanan
Typeset in Minion by Tseng
Information Systems, Inc.
Library of Congress Cataloging-in-
Publication Data appear on the last
printed page of this book.

To the memory of
Clifford Geertz (1926–2006)
colleague and friend

CONTENTS

Introduction

ANGELA N. H. CREAGER, ELIZABETH LUNBECK,
AND M. NORTON WISE

At the dawn of the twenty-first century, the face of biology may well be that of a laboratory mouse. Science writers, government agencies, and researchers alike tout the crucial role played by biology's experimental subjects, "model systems" as they are termed, in advancing knowledge. These creatures are not showcased for their appeal—the flies, mice, worms, and microbes that are the mainstay of laboratory science would be regarded as vermin or germs outside their scientific homes—but because they have become the locus of producing knowledge about life and disease. To make the case that improving human health rests on our intimate understanding of a select set of rodents, fish, amphibians, microbes, and even a plant, the National Institutes of Health (NIH) features a Web site titled "Model Organisms for Biomedical Research" (www.nih.gov/science/models). These are the organisms whose genomes were sequenced as part of the Human Genome Project. And as the NIH wants to make clear, they are the creatures that stand in for us humans as laboratory biologists investigate how living processes work—and how they go awry. A special supplement to *The Scientist* titled "Model Organisms" offers feature articles on eight such exemplary forms of life, from the intestinal bacterium *Escherichia coli* to the nematode worm *Caenorahdbitis elegans* (see figure 1). As the editors explain the importance of this "motley collection of creatures":

> Researchers selected this weird and wonderful assortment from tens of millions of possibilities because they have common attributes as well as unique characteristics. They're practical: A model must be cheap and plentiful; be inexpensive to house; be straightforward to propagate; have short gestation periods that produce large numbers of offspring; be easy to manipulate in the lab; and boast a fairly small and (relatively) uncomplicated genome. This type of tractability is a feature of all well-used models.[1]

At one level, the reliance of biomedical researchers on standardized creatures for experimentation is mere practical necessity. Biological materials are, by their nature, variable and complex; life scientists have sought to control the variability they face by selecting out and standardizing particular experimental subjects. Yet these organisms, no matter how standardized they become as laboratory instruments, maintain an independent existence in a contingent world. They are not models in the traditional sense—they are not smaller versions of humans, and they do not exactly replicate our experiences or diseases. Unlike the idealized representational models characteristically featured in the history of the exact sciences, in which the model (e.g., the Bohr atom) has been supposed to mirror a natural system (hydrogen) by embodying the mathematical laws and structure from which the behavior of the system can be deduced, model systems maintain their own autonomy and specificity. That is, model systems do not directly represent humans as models *of* them. Rather, they serve as exemplars or analogues that are probed and manipulated in the search for generic (and genetic) relationships. They serve as models *for* human attributes.[2] The use of standardized organisms in biomedicine is part of a broader model-systems approach in the life sciences that includes the investigation of a far wider range of entities, from specific proteins (e.g., hemoglobin) to particular lakes (e.g., Linsley Pond in Connecticut), and whose utility in producing general knowledge relies on the routine use of analogies to other examples and entities.

These distinctions between representational and representative functions, between models of and models for, have proven quite useful in discussing the characteristics of model systems. We suggest that insofar as similar objects inhabit spaces far beyond biology laboratories, the same distinctions extend to other areas, areas where relations of similarity rather than deduction have grounded claims to generality and where specificity has been a resource rather than a problem. Many fields have developed canonical examples that have played something like the role of model systems, which serve not only as points of reference and as illustrations of general principles or values but also as sites of continued investigation and reinterpretation. What we here call model objects of this sort in this volume include Athenian democracy in political theory, the ritual in anthropology, and the so-called Prisoner's Dilemma in game theory. Through what processes do particular organisms, cases, materials, or texts become foundational to their fields? How do they serve a classificatory function for the organization of knowledge, whether it is in a biology laboratory or an art museum? When does the specificity or idiosyncrasy of an example threaten its utility?

Examining the pursuit of knowledge organized around exemplars rather than around fundamental laws, we aim to reopen the old question of the relation between the human sciences and the natural sciences. In the nineteenth century, the question was cast in terms of the relation between the generalizing,

FIGURE I Cover Image, "Model Organisms," *The Scientist* 17, supplement 1 (2003). Reprinted with permission of *The Scientist*.

lawlike sciences (*nomothetic*, in the canonical formulation of Wilhelm Windelband) and the particularizing sciences (*idiographic*), where lawlike referred to the universal laws of physics as the ideal of science.[3] It is no longer the case, however, that universal laws either do or can serve as a model for all science, even natural science. This has become most apparent with the emergence of biology in the past thirty years as the so-called science of the future. It is not clear that there are any high-level laws in biology, in the sense of predictive laws that determine the future behavior of a biological system (except perhaps in evolutionary theory); we will not be concerned with whether such laws may emerge. Instead, we want to show how the model-systems approach so pervasive in biology compares with the use of cases, exemplars, and related methods in other fields. Interestingly, it appears that many of these approaches grew up in response to the challenge of producing something like lawlike knowledge in disciplines in which laws seemed incapable of capturing the specificity and complexity of organisms, geological processes, or human productions. If the result has not been laws, it has nevertheless been reliable, systematic knowledge. Thus our title: *Science without Laws*.

Beginning with model systems as understood in recent biology, then, this book compares the scope and function of model objects in domains as diverse as geology and history, attending to differences between fields as well as to epistemological commonalities. What distinguishes this collection of essays from other studies of models in science is both its attention to model systems as concrete autonomous objects and the breadth of disciplines it addresses.[4] Traditional approaches to scientific models derive from the philosophy of science, especially the philosophy of theoretical physics, in which the model is presented as a precise, usually mathematical, representation of the phenomenon in question.[5] But this picture of mathematical models, valorized by physicists and philosophers, is highly idealized. As Nancy Cartwright has argued, the models mobilized by physical scientists to illustrate their theories are far from real-world situations; they are, rather, "nomological machines" that manufacture universal laws of nature by providing the kind of simplified mechanical or mathematical evidence that could not be found in "nature."[6] Not only the character of the model but also the presumption that the exact sciences should provide the basis for understanding general scientific method or rationality has recently come under question. To Ian Hacking's enumeration of six "styles of reasoning" that characterize the sciences, for example, John Forrester has proposed "reasoning by cases" as a seventh scientific method, widely used not only in the human (and biological) sciences but also in law, medicine, and ethics.[7] Case-based reasoning relies on relations of similarity rather than on conventional reductionism and treats specificity as a resource, not a problem. The essays in this book attend to case-based modes of inquiry usually neglected by historians and philosophers of science, demonstrating that their epistemological practices and patterns extend far beyond the boundaries of "science."

PART I: BIOLOGY

A model system in biology refers to an organism, object, or process selected for intensive research as an exemplar of a widely observed feature of life (or disease). The traditional contrast between laboratory physiology and natural history within nineteenth-century biology provides some background to the emergence of model systems in twentieth-century laboratories. Of the two traditions, it was experimental physiology that most closely emulated the ideals of the physical sciences. Nineteenth-century physiologists used animal models such as frogs because of their accessibility for experimental manipulation, and they aimed to produce universalistic knowledge through reductionist (and often instrument-based) approaches. This mechanistic approach treated the organism itself as a workshop, but there was considerable debate as to whether vital phenomena could be completely reduced to physico-chemical principles or laws. Hermann Helmholtz, Emil Du Bois-Reymond, Carl Ludwig, and others strove to reduce physiology to the physics of atoms and forces in their attempts to show how laws (such as the conservation of energy) might account for processes in both living and nonliving materials. Not all physiologists agreed with the aims of this "organic physics," however. Claude Bernard cautioned that biology might borrow methods and instruments, but not theory, from physics and chemistry. In fact, radical reduction soon went bankrupt and even its most committed adherents had to redirect their energies. Ludwig, who in the early 1850s had asserted that "physiology is nothing other than applied physics," had also at the same time expressed his "hope someday to work with a capable clinician or pathological anatomist . . . and together with him experimentally reproduce the conditions of disease. . . . It ought to be possible to generate innumerable illnesses similar to those found in man." This is the expansive program he took up with Karl Wunderlich when he moved to Leipzig in 1871 to establish the new physiology laboratory there.[8] It is also the program that turned increasingly to animal models rather than physical laws for insight into biological processes.

The turn to model systems in the twentieth century resulted from the conjunction of this experimental tradition with a new industrial infrastructure. Accompanying the increasing mass production of scientific materials and equipment was a narrowing of the number of organisms intensively studied, and also the commodification of many of the laboratory's experimental inhabitants. The ascendance of the fruit fly in genetics research grew out of T. H. Morgan's laboratory in the 1910s and 1920s; during the same time period, maize became a dominant organism for plant genetics, particularly cytogenetics. By midcentury, inbred strains of mice were widely used not only to understand mammalian genetics but also for investigations of cancer and other human afflictions. And the molecular emphasis of postwar biology grew out of the intensive study of a handful of microbes, especially yeast (*Saccharomyces cerevissiae*), bacteria (most notably *E. coli*), and viruses (particularly the bacteriophages and tobacco mosaic virus). Since the 1960s, scientists have domesticated new animals for research, such as the nematode worm and the zebrafish. Each of these model organisms has a unique history in the laboratory, with particular physical features and experimental advantages. Yet each of these model systems has also gained ground by virtue of its historically acquired prominence within a field of study. Indeed, model systems exhibit a self-reinforcing quality: the more the model system is studied, and the greater the number of perspectives from which it is understood, the more it becomes established as a model system. Even for the many biologists who do not study one of the canonical model organisms, these systems tend to serve as benchmarks and methodological guides when they turn to other organisms and objects as researchers.[9]

Historians and philosophers of biology have recently become interested in the consequences of privileging particular model systems for the shape of knowledge. In genetics, as Marcel Weber shows, the collection of mutant *Drosophila* strains that dated to Thomas Hunt Morgan's "fly room" in the early twentieth century were redeployed in the molecular mapping of fly genes in the 1970s. The recombinant revolution in biology thus catalyzed the reworking of a classic model system, but one that was advantaged by the accumulation of decades of knowledge and hands-on experience, not to mention the fly strains themselves. *Drosophila* was especially ascendant in the field of developmental genetics, and there its success cannot be attributed to a simple notion of typicality. The process of development in *Drosophila* is not characteristic of other organisms, especially mammals. Indeed, some biologists argue that the highly canalized development of *Drosophila* that makes its development so easy to study in the laboratory also makes it a relatively poor evolutionary representative of its own phylum.[10] Weber argues that what made *Drosophila* such a powerful research tool were rather the experimental resources associated with it. In the process of its molecularization, he argues, key concepts from classical genetics—not least the concept of the gene itself—were mobilized and recast.

For a time, the combination of accumulated knowledge and resources associated with the fly gave it an edge over other experimental organisms, such as the mouse and the worm, in efforts to adapt cloning techniques from the study of microbes to multicellular organisms.

Not all current model systems are advantaged by such a long history in the laboratory. Rachel Ankeny describes the way in which a newly developed model system, the nematode *Caenorhabditis elegans*, came to figure very prominently in medical research following its initial domestication in the 1960s. Sydney Brenner, already well known for his work in molecular biology, chose this free-living nematode on the basis of its developmental invariance and simplicity in order to facilitate developmental studies of the nervous system. As Ankeny argues, this kind of modeling of a general process on a carefully selected example involves at least two levels of idealization. First, the so-called wild type that serves as a benchmark for genetic comparison may not be especially representative in terms of the variation within a species, but it serves as a biological norm nonetheless. Second, a great deal of scientific work goes into the building up of a descriptive model of the organism in question. In the case of *C. elegans*, this work centered on the articulation of a wiring diagram that represented all of the neurological connections within the worm. Once again, this diagram is itself an idealized version of actual worms, whose properties are abstracted into the diagram. Even so, it provides a crucial point of reference for the analysis of various mutants. As Ankeny demonstrates, a well-studied model organism (or at least its wild-type representative) serves as an index case against which others can be compared and contrasted. In this respect, case-based reasoning underlies the routine use of model systems in biomedical research, particularly when properties of genetic mutants are compared against the wild type.

Analogies also play a crucial role in linking worms and humans, particularly at the level of homologies between genes. It is these analogies that have given nematode worms their purchase in medical research. This process is best described in terms of an unpredictable relevance, as Jane Hubbard puts it, for genetic homologies that can underlie divergent physiological properties. The molecular biology of organisms such as the worm began as unabashedly reductionistic in spirit, yet the specificity and particular features of the model systems have remained experimental resources rather than mere complications. And although model organisms are standardized in order to facilitate highly controlled biological experimentation, their inherent complexity means that the systems are never fully understood and can continue to generate surprising results. Indeed, as models, they are no simpler biologically than the humans they illuminate by analogy.

The career and utility of a model system not only derives from pragmatic experimental considerations but may also reflect broad disciplinary and political changes. Susan Sperling shows that the intensive study of the baboon in field primatology of the 1970s coincided with the declining legitimacy in the anthropological study of "primitive peoples" as stand-ins for our prehistoric ancestors. The structural-functionalist anthropologists in the 1960s were no less universalizing than their disciplinary predecessors, but they turned from anthropology to primatology in their efforts to theorize human evolution. As Sperling observes, "the ubiquitous baboon troop now firmly occupied the position of primitive society in the schemas of nineteenth-century anthropologists." Postulating that the earliest humans behaved like baboons, with their patterns of male dominance and troop control through aggression, primatologists provided a putative biological basis for similar human behaviors. These visions of the human past and present did not go uncontested, as a generation of new researchers, many of them women, found the received view of baboons' gendered division of labor inadequate to describe troop behavior. In addition, a new emphasis on other primate model systems, such as chimpanzees and lemurs, complicated the vision of an ur-primate society based principally on baboon observations.

PART 2: SIMULATIONS

As we leave the domain of biological organisms and contemplate extensions of the concept of a model system into other areas, it is helpful to return again to the mid-nineteenth century. As noted previously, anyone at that point seeking a contrast with the sciences of laws would immediately have turned to natural history in opposition to natural philosophy (or physics, as the field was called after midcentury). Natural historical sciences like botany, zoology, geology, and meteorology typically relied for their claims to knowledge on extensive observation rather than experimentation, and on the classification and ordering of large amounts of quite specific information rather than on subsuming particulars under general laws. The understanding of a generic trait of an organism, for example, depended on knowing what was typical and what was possible within the range of variation for that organism and on a thorough familiarity with the analogues and homologues of the trait in other organisms. Similarly, through extensive fieldwork and mapping projects on rocks and fossils, geologists of the early nineteenth century (geognosists) made great strides in understanding the order and temporal sequence of rock strata.[11] Such descriptive ordering, however, did not yield causal laws. When investigators from the natural historical sciences wanted to understand causes—thereby taking on the usual goal of natural philosophy—they sometimes had recourse to models, but as Naomi Oreskes shows for geology, they were models of a distinctly natural historical sort, based on mimesis rather than on general laws of cause and effect. Such mimetic models may be thought of both as *analogues* of the real system, being like it, and as *analog* simulations of its development in time. This suggests an interesting similarity to present-day digital simulations, with which both Oreskes and Amy Dahan Dalmedico are concerned. Their studies suggest that many forms of computer simulation, in areas in which laws are either inappropriate or imperfectly understood, may be usefully interpreted as continuations of natural historical methods of modeling.[12]

Oreskes makes this argument directly for the use of models in the earth sciences. She describes the development of scale models, from the early nineteenth century to the mid-twentieth century, for investigating such things as the deformation of the earth's crust and mountain building associated with the common assumption that the earth was a cooling ball of originally molten matter. A long line of geologists constructed "boxes" containing model materials for the earth's crust that could be subjected to compression and other forces of deformation in order to imitate, or to run an analog simulation of, geological processes. The difficulty lay in finding materials for the models—lead, sandstone, paraffin, or pancake batter—that would behave in the small-scale box on a short time-scale in a manner similar to the large-scale plasticity of the earth's rocks on a much larger time scale. Characteristic of these models was their heuristic role; they allowed experimental investigation, by analogy, of possible causal stories whose plausibility might thereby be supported. And they allowed the investigator to watch the story developing dynamically, in time.

Interestingly, by the time geologists had developed effective rules for the scaling-up of materials and forces from the box to the landscape, they realized that they could just as well calculate the results as build and run the physical model. With this realization in the 1940s, and with the advent of digital computers, they largely gave up their analog models in favor of running digital simulations. The computer simulations, however, while substituting digital plasticity for the plasticity of model materials, continue in some ways to serve much the same exploratory and explanatory function as the analog models. They map out the possible while seeking understanding in the details of carefully chosen objects whose very specificity constitutes the source of insight. Thus both sorts of models, analog and digital, might well be called the model systems of the earth sciences. The new simulations, however, have taken on a role foreign to geology's natural historical origins, that of predicting future events. Oreskes argues that this dramatic shift in the modeling goals of geologists, from an exclusive concern with explanation to an increasing emphasis on prediction, derived originally from the Cold War problem of predicting the long-term stability of disposal sites for nuclear waste generated by a mammoth weapons program. Long-term prediction using computer simulations presented new sorts of problems for geological modeling, including the validation of results for a distant future not accessible to observation.

More generally, the computer simulations that Oreskes discusses contain a feature of model systems that we have not yet explicitly addressed: their complexity. Like biological model systems, analog and digital simulations are of immense value precisely because they mimic in part the complexity of natural systems, which typically involve multiple processes, nonlinear interactions, feedback loops, and emergent properties. These characteristics make complex systems irreducible in mechanical terms and thus impossible to replicate, except in a similarly complex analogue. It might well be said, in retrospect, that the traditional methods of natural history for producing order out of nature's diversity, including the role of analogues, were already one sort of response to the problem of complexity. Model systems and computer simulations are another. They are extended forms of analogues, but they offer the great advantage of experimental exploration of those aspects that produce complexity. In computer simulations, those aspects are built into the models and modified to match some of the features of nature's own behavior.

Amy Dahan Dalmedico analyzes the practices that have gone into building and modifying meteorological and climatological models since the 1950s, particularly in France. Like Oreskes, she finds that prediction, or forecasting, presents problems different from those of understanding mechanisms and causes. In fact, the inherent tension between predicting and understanding has constantly appeared not only within the models but between modelers from different institutions with different interests. For climate change especially, the tension is greatly enhanced by the interdependence of models of the natural biosphere, models of the effects of various scenarios for socioeconomic development, and models for possible mitigation strategies. A full climate model must integrate these partial models and the feedback between them. The complexity of the full model, Dahan Dalmedico shows, mirrors the complexity of these interactions and the various modeling practices associated with them. As a whole, they constitute the model system of climatology, a science both natural and social.

To illuminate the different modeling practices that must somehow be integrated into a global model, Dahan Dalmedico compares the two main groups in France that model climate change. Météo-France is a national public institution whose approach derives from meteorological forecasting and which uses a large computer simulation called the Arpège model. The staff comprises primarily engineers who become civil servants and are interested in the operational aspects of modeling. The laboratories of the Institut Laplace, in contrast, are supported by the CNRS (National Center of Scientific Research, somewhat like the American National Science Foundation) and are primarily interested in basic research on the components of models. The organization of the work is relatively loose, and individual members work on a variety of topics. One of the laboratories, anxious not to lose their independence to Météo-France and the Arpège model, built their own large climate model, called LMDZ, which is not immediately compatible with Arpège. Neither model, furthermore, takes into account the coupling of the atmosphere with the ocean. For that purpose, a third organization has arisen, Cerfacs, whose coupling model attempts to maintain neutrality between Arpège and LMDZ, taking the results of either as the input for higher-order integration. Yet further levels involve the continents, in both soil properties and biosphere.

The result of all of this integration is a virtual climate so complex that it constitutes an almost autonomous object, a "simulacrum of reality," as one researcher called it, whose self-consistency sometimes seems to compete with reality. This worry reflects again the tension between understanding and predicting. But in their stubborn autonomy, climate models recall a basic characteristic of model systems that we have wanted to stress. They do not represent the climate in the sense of being models *of* it; instead, their properties represent certain aspects (carbon dioxide content, temperature) and are models *for* those aspects. They yield understanding by revealing processes in the model that are closely analogous to those in the atmosphere, although realized in a different medium, the computer simulation. That is just the characteristic, at a completely different level, that seems typical of the carefully selected strains of mice that are used to investigate the genetics of particular disease entities. Of course a mouse is not a computer model, and a mouse cannot be abstracted from its life-support system; still, in both these cases of modeling complex systems, to understand a thing is to understand the workings of a model *for* it rather than a model *of* it. Along these lines, Dahan Dalmedico cites the apt remark of one of the originators of meteorological simulation, John von Neumann, who said that "the sciences do not try to explain . . . they mainly make models."[13]

Von Neumann went on to define a (mathematical) model as "a mathematical construct which, with the addition of certain verbal interpretations, describes observed phenomena."[14] Conscious that these "certain verbal interpretations" played a role, he nevertheless did not make much of them. Mary Morgan takes up the missing link in her discussion of how games derived from the Prisoner's Dilemma have become the *E. coli* of economics. Economists have long sought after a science of laws governing general equilibrium in a perfectly competitive market in which individual actors are utility-maximizing rationalists, but the Prisoner's Dilemma, Morgan shows, has rather insidiously undermined their faith. It became a pervasive means of conceptualizing economic action during the Cold War, after game theory was introduced at the RAND Corporation and the Institute for Advanced Study. The Cold War, in the eyes of analysts, came simply to be a sequence of prisoner's dilemmas.

Morgan wants us to appreciate just how bizarre it is that the wide world of international diplomacy got squeezed into the 2 x 2 matrix of payoffs for the Prisoner's Dilemma. She would have us bring our incredulity to the question of how economists could have developed the same little matrix into a proliferating set of games for reasoning about complicated economic situations. How could such an apparently poverty-stricken tool prove so versatile? Morgan's answer breathes life into von Neumann's "certain verbal phenomena." She argues that it is the narratives accompanying the matrix that adapt it to various situations by spawning new games, for it is the text, not the matrix, that contains both economists' traditional assumptions about rationality and equilibrium and the rules of the game to be played. Thus the narratives produce quite specific types of games as model situations appropriate for exploring particular kinds of economic behavior. All of this particularity in the narratives suggests that they might well be compared to certain modes of analysis common in the human sciences. We take up two of those methods in our concluding section.

PART 3: HUMAN SCIENCES

Focused methodological engagement with the problem of extracting the universal from the particular in the human sciences first occurred in the late nineteenth century, when newly emerging fields like sociology and anthropology sought professional identification and status. The physical sciences themselves had only recently gained credibility in their aim to subject the formerly experimental subjects of electricity, magnetism, and heat to general mathematical laws and to unite all branches of physics under the laws of energy conservation and entropy increase. Evolutionary theory in biology and newly discovered statistical regularities in society promised equivalent generality. To many, the time seemed ripe to attempt something similar for more directly human productions. We are most familiar with the problem in its negative form, articulated by Windelband as the contrast between the methods of the *Geisteswissen-schaften* and the *Naturwissenschaften*, and by Wilhelm Dilthey as the contrast between the requirements of understanding (*Verstehen*) and those of explanation (*Erklärung*).[15] Closely bound up with that problematic are two interrelated modes of dealing with it, the case and the exemplary narrative.

Two of our authors have written elsewhere on cases and exemplary narratives in the human sciences. John Forrester, in his 1996 article, "If *p*, Then What? Thinking in Cases,"[16] argued that the scientificity of psychoanalysis—and, for that matter, the human sciences in general—was decidedly nonuniversalistic, organized principally not around the attainment of generalizable laws, but rather on knowledge of what Aristotle called "infinitely various" individuals, or cases. The Aristotelian position holds that there can only be sciences of what is universal and generalizable, but, as Forrester shows, the late nineteenth century saw the proliferation of case-based reports and of sciences—psychiatry, psychology, criminology, and the other clinical disciplines—premised on such reports and their capacity to render individual lives in a form amenable to the conventions of science. Since then—in the human sciences and in law and medicine—"the case" has functioned much like a model system. Cases in these disciplines are valued for their specific material reality, their individuality, and at the same time their typicality. Cases, it is assumed, capture individuals in all their complex uniqueness while at the same time rendering them in a generically analyzable form. The disciplines of psychiatry and psychoanalysis in particular have been constructed over the course of the twentieth century largely from the knowledge of cases, both those published in journals and books and those discussed and passed down as part of the field's oral traditions.

Case thinking found its first formal disciplinary instantiation as a system of instruction, introduced in the 1870s by C. C. Langdell of the Harvard Law School, who promoted it as a means by which aspiring lawyers could master the law's principles and governing doctrines through the study of "the cases in which it [was] embodied."[17] Walter B. Cannon and Richard Cabot soon after imported the case method to Harvard's medical school and teaching hospitals, and, in 1911, the university's newly established business school followed suit. Forrester points out that in law in particular, but in the clinical sciences as well, a strong tradition of reasoning by example, of, in Oliver Wendell Holmes's words, "reasoning from case to case," has profoundly but indeterminately shaped practices and that it has in turn found renewed life in the new casuistic science of bioethics.[18] Yet however durable the science of the particular fact has proven, a tension between an interest in capturing complexity and individuality and the demand for system, for formal principles—often presented as the demand of "science"—has historically run through the case-based disciplines, generating epistemological confusion and practical controversies. Can the case serve both for generating rational principles and for maintaining locality and specificity? What has been the history of this relation? The essays by Forrester and Clifford Geertz explore aspects of the ways in which cases function like model systems in such respects and how certain cases have achieved exemplary status within their disciplines, serving as nodal points around which practice—including teaching, research, and the generation of theory—have been organized.

Forrester's reflections on the consolidation of a new epistemological model that fixed on specificity rather than generality in the late-nineteenth-century human sciences complement the historiographical analyses of Carlo Ginzburg. In his now classic 1983 article, "Clues: Morelli, Freud, and Sherlock Holmes," Ginzburg traced the emergence of what he called the conjectural or semiotic model, counterposed to Galilean science's stress on measurement and experiment, that focused on the particular and on individual cases and that favored "little insights" over grand theory, traces and clues over systematic knowledge. The more a discipline was concerned with individuals, Ginzburg wrote, "the more difficult it became to construct a body of rigorously scientific knowledge."[19] The human sciences, concerned as they were with the individuals and their particularities, perforce accepted the conjectural paradigm, which prefigures Forrester's case thinking, characterized by reasoning "from particulars to particulars." History is a case in point, "irremediably based in the concrete," Ginzburg writes: "Historians cannot refrain from referring back (explicitly or by implication) to comparable series of phenomena; but their strategy for finding things out, as well as their expressive codes, is basically about particular cases, whether concerning individuals, or social groups, or whole societies."[20] They seek the universal through the particular, through the clues of the highly specific and located—the basis of Ginzburg's own microhistorical method. While the medically trained art historian Giovanni Morelli offers the most direct historical source for the method of clues, Sherlock Holmes remains the most colorful representative of both the method and its origins in medical diagnosis (through the training of Sir Arthur Conan Doyle). Holmes cared nothing for abstract laws of nature or for such grand schemes as the Copernican system. Instead he used his profound knowledge of the natural historical aspects of sciences like chemistry and geology, and of mundane things like tobacco ash and mud, to connect one particular case to another by analogy. "I am generally able, by the help of my knowledge of the history of crime, to set them [confused colleagues] straight. There is a strong family resemblance about misdeeds, and if you have all the details of a thousand at your finger ends, it is odd if you can't unravel the thousand and first."[21] Although Holmes did not aim at general principles, his method of clues does not differ as much as one might expect from views sometimes expressed by mathematicians like von Neumann and Stanislav Ulam about the heuristic use of the computer to obtain insights into pure mathematics. Ulam remarked: "By producing examples and by observing the properties of special mathematical objects, one could hope to obtain clues as to the behavior of general statements which have been tested on examples."[22]

Ginzburg points out that Freud found inspiration in the Italian physician Morelli's method, writing that his method of inquiry was "closely related to the technique of psychoanalysis"—divining "secret and concealed things from despised or unnoticed features."[23] In psychoanalysis, as in the other sciences of humans, the interpretation of clues formed the substance of the science. Similarly, Forrester's essay here on Robert Stoller's classic case study, *Sexual Excitement*, invites readers to explore the question of how the case figures in psychoanalysis. But rather than constitute a formal inquiry into the issue, this essay is itself distinctively psychoanalytic, addressing theoretical questions through the mining of one exemplary case. Forrester highlights the distinctive function of the psychoanalytic case, suggesting that it attempts to convey the experience of both analyst and patient, eschewing any claims to disinterested scientificity. He argues that the case, as a genre of writing, replicates the transferential and countertransferential relations that govern the analytic situation; it cannot transcend the conditions of the psychoanalytic encounter it is meant to document. The actual writing of the case thus enacts the very forces it attempts to capture—exhibitionism, in this instance, which figures variously as Stoller's patient's signal symptom, as a charge against which Stoller preemptively defends himself in an endlessly reiterated and deferred way, and, finally, as a charge against which Forrester, in turn, must mount his own defense. But both of these defenses are bound to fail, he suggests, for not only does the narrative force of this specific case prompt one at every turn to assume the position of the voyeur one disavows but the psychoanalytic case as a genre traduces the condition of confidentiality that is at the heart of the analytic transaction, inviting the reader to, in Stoller's words, peek in on forbidden scenes. Exhibitionism is thus a formal characteristic of the case in its written, transmissible form. Readers of the case necessarily participate in the perversion of looking in on forbidden scenes; in transmitting these scenes, Forrester both participates in the voyeurism and attempts, through his exhaustive analysis, to mimic the depletion of transferential eroticism that Stoller informs readers was a condition of publication. Forrester's essay consists in a layered enactment of and reflection on the conditions under which psychoanalytic knowledge is produced and transmitted. Can writers on psychoanalysis transmit psychoanalytic knowledge, or are they fated, like the analyst, to "infect" readers with its terms and frames of reference?

Clifford Geertz's essay on rituals as model systems in anthropology examines another facet of the human sciences. Surveying the deployment of ritual in generations of classic anthropological texts (or case studies), Geertz reconsiders both functionalist and structuralist approaches before reframing the question of what it is rituals model in hermeneutical terms: rituals, he proposes, model the cultural achievement of having confidence in the "reality" of one's world, its depth and substantiality, and, at the same time, the threat of losing that confidence. Rituals—characterized here as "cultural *Drosophila*," like Morgan's Prisoner's Dilemma as the *E. coli* of economics—thus not only model practical action and category change but bring to light modes of being-in-the-world. Geertz traces the historical fate of ritual, from, in anthropology's early days, an object of study to, more recently, an object to study with: "the microscope, not the bug under it." He concludes that ritual might usefully be considered akin to model systems in biology, sharing with them the characteristics of specificity, typicality, materiality, and complexity. Further, he argues, ritual can model just about anything that the anthropologist finds in need of explanation. Assuming the position of observer of moderns' lostness and alienation in the world—pointedly evident in the schizophrenic's apprehension of her surroundings as strange and peculiar, herself as detached and outside—Geertz provocatively concludes that what is missing, a commonsense orientation to reality, is precisely what ritual provides, not only to the individual in a culture but also to the anthropologist studying that culture.

It will be apparent that canonical case studies of ritual, such as Geertz's of the Balinese cockfight, and exemplary psychoanalytic cases, like Freud's of Dora, depend for their effectiveness on the specificity of the narrative that expresses them. In this they share something with the Prisoner's Dilemma, whose applicability depends on a narrative construction. Arguably, the case, even when highly standardized to resemble a laboratory mouse, depends crucially on its accompanying narrative, for it is the narrative that conveys its specificity. This suggestion obtains further support from the article by Josiah Ober on Athenian democracy as an exemplary narrative for normative political theory. Ober faces the alternative charges of historicism and essentialism leveled, respectively, against those (historians) who would prioritize specific instances of democratic organization and against those (political theorists) who would propose transcendent universal principles. He finds a middle ground in which the Athenian case, as a richly articulated narrative of political practice, can serve as the basis for running thought experiments on such things as rights, citizenship, and civic virtue. It can help to frame the questions and expand the scope of contemporary ethical intuitions while challenging the capacity of any universal laws of democratic development to deal fully with historically specific political formations. Democratic Athens, as a model system, enriches political theory, while political theory enriches the model system of democratic Athens.

We end this volume with Carlo Ginzburg's self-proclaimedly experimental microhistory, in which he directly addresses the question that animates this section: "Can an individual case, if explored in depth, be theoretically relevant?" Answering his question in the affirmative, Ginzburg shows how plumbing the particular is not to abjure generalities, but to open the possibility of unsettling them—in this instance, those that inform Max Weber's ideal-typical rendering of *The Protestant Ethic and the Spirit of Capitalism* and Karl Marx's earlier alternative in *Capital*. Ginzburg is no proponent of the inductive methodology championed by early-twentieth-century theoreticians of the case examined by Forrester. Rather, taking his cue from Weber's argument that ideal types—representing Weber's attempt to capture human universals—must be continuously tested, he uses the obscure but detailed story of an actual Calvinist entrepreneur of the early eighteenth century to critique the missing role of violence and conquest in Weber's thesis and the missing religious agents in Marx. Ginzburg's entrepreneur, Jean-Pierre Purry, turns out to be altogether incompatible with the Weberian ideal type of the ascetic early capitalist forerunner, which forces not so much a revision of the ideal type as a reconsideration of the theoretical utility of such a construct. Ideal types, by Weber's telling, are "conceptual wholes," distillations of characteristics from innumerable individual examples. Cases and exemplary narratives, organized around specificity and particularity, are resistant to the level of generality on which the ideal type is premised; Purry is too idiosyncratic a figure to fit Weber's scheme. But neither is his case to be the occasion for qualifications and tinkering at the margins. Rather, it prompts Ginzburg to lay out the theoretical possibilities of the particular, to question the reflexive relegation of the individual case to the periphery of theory and to mount instead an argument for its centrality to historical writing, its capacity to yield interpretive riches comparable to—even exceeding—those of theory. Ginzburg closes his piece on a Proustian note in which the problematic of this volume is nicely anticipated. "People foolishly imagine that the broad generalities of social phenomena afford an excellent opportunity to penetrate further into the human soul; they ought, on the contrary, to realise that it is by plumbing the depths of a single personality that they might have a chance of understanding those phenomena."[24]

NOTES

This volume had its origins in a seven-session workshop, convened by the editors, "Model Systems, Cases, and Exemplary Narratives," in the Program in History of Science at Princeton University, 1999–2001. We thank the history department and Shelby Cullom Davis Center for supporting these events and graduate students Renée Raphael and Doogab Yi for assistance in preparing the manuscript. Two individuals outside the program faithfully attended nearly all the workshops, keeping a continuous conversation going in addition to contributing their own papers and comments: Mary Morgan and Clifford Geertz. We were deeply saddened that Cliff passed away while the volume was in production, and dedicate this book to his memory.

1. Christine Bahls, Jonathan Weitzman, and Richard Gallagher, "Biology's Models," in "Model Organisms," *The Scientist* 17, supplement 1 (2003), 5.

2. The distinction made here calls upon a more fine-grained understanding of how models function in various different representing roles compared with the description sometimes stressed in philosophy of science that scientists build models *of* something *for* some purpose. For an account that does pay close attention to how models represent within that latter framework, see Evelyn Fox Keller, "Models of and Models for: Theory and Practice in Contemporary Biology," *Philosophy of Science* 67 (2000): S72–S86. We are indebted to Mary Morgan for bringing the model of/model for distinction to our attention, and helping clarify its role in our analysis.

3. Wilhelm Windelband, "Geschichte und Naturwissenschaft," in *Präludien: Aufsätze und Reden zur Philosophie und ihrer Geschichte* (Tübingen, Germany: Mohr, 1915): 145; reprinted as "Rectorial Address, Strasbourg, 1894" (trans. Guy Oakes) in *History and Theory* 19 (1980): 169–85.

4. Two other collections highlight the role of concrete objects in the formation of knowledge: Soraya de Chadarevian and Nick Hopwood, eds., *Models: The Third Dimension of Science* (Stanford: Stanford University Press, 2004), examines the role of three-dimensional models in science, technology, and medicine. Lorraine Daston, ed., *Things That Talk: Object Lessons from Art and Science* (New York: Zone, 2004), considers more broadly how particular made "things" articulate what we come to know.

5. See, for example, Mary B. Hesse, *Models and Analogies in Science* (London: Sheed and Ward, 1963). We have taken inspiration from a recent reconsideration of scientific models: Mary S. Morgan and Margaret Morrison, eds., *Models as Mediators: Perspectives on Natural and Social Science* (Cambridge: Cambridge University Press, 1999). Although the models Morgan and Morrison address are by and large mathematical, not material, their emphasis on the partial autonomy of models connects with our interest in the autonomy of model systems.

6. Nancy Cartwright, "Nomological Machines and the Laws They Produce," in *The*

Dappled World: A Study of the Boundaries of Science (Cambridge: Cambridge University Press, 1999), 49–74.

7. John Forrester, "If *p*, Then What? Thinking in Cases," *History of the Human Sciences* 9 (1996): 1–25. Ian Hacking, *The Taming of Chance* (Cambridge: Cambridge University Press, 1990), 6–7; following A. C. Crombie, "Philosophical Presuppositions and Shifting Interpretations of Galileo," in *Theory Change, Ancient Axiomatics, and Galileo's Methodology*, ed. Jaakko Hintikka, David Gruender, and Evandro Agazzi (Dordrecht: Reidel, 1981), 271–86. Hacking developed the notion at more length in "Styles of Reasoning," in *Post-Analytic Philosophy*, ed. John Rajchman and Cornel West (New York: Columbia University Press, 1985), 145–64.

8. Timothy Lenoir, "Science for the Clinic: Science Policy and the Formation of Carl Ludwig's Institute in Leipzig," in *Instituting Science: The Cultural Production of Scientific Disciplines* (Stanford: Stanford University Press, 1997), 113, 127. See also Frederic L. Holmes, "The Old Martyr of Science: The Frog in Experimental Physiology," *Journal of the History of Biology* 26 (1993): 311–28; Frederic L. Holmes, *Claude Bernard and Animal Chemistry: The Emergence of a Scientist* (Cambridge: Harvard University Press, 1974), 1–2; Claude Bernard, *Introduction à l'étude de la médecine expérimentale*, 3d ed., trans. Henry Copley Greene (New York: Dover, 1957), 149–50; Robert M. Brain and M. Norton Wise, "Muscles and Engines: Indicator Diagrams and Helmholtz's Graphical Methods," in *Universalgenie Helmholtz: Rückblick nach 100 Jahren*, ed. Lorenz Krüger (Berlin: Akademie Verlag, 1994), 124–45.

9. Robert E. Kohler, *Lords of the Fly: Drosophila Genetics and the Experimental Life* (Chicago: University of Chicago Press, 1994); Nathaniel C. Comfort, *The Tangled Field: Barbara McClintock's Search for the Patterns of Genetic Control* (Cambridge: Harvard University Press, 2001); Karen A. Rader, *Making Mice: Standardizing Animals for American Biomedical Research, 1900–1955* (Princeton: Princeton University Press, 2004); Ilana Löwy and Jean-Paul Gaudillière, "Disciplining Cancer: Mice and the Practice of Genetic Purity," in *The Invisible Industrialist: Manufactures and the Production of Scientific Knowledge*, ed. Gaudillière and Löwy (London: Macmillan, 1998), 209–49; Joshua Lederberg, "*Escherichia coli*," in *Instruments of Science: An Historical Encyclopedia*, ed. Robert Bud and Deborah Jean Warner (New York: Garland, 1998), 230–32; Angela N. H. Creager, *The Life of a Virus: Tobacco Mosaic Virus as an Experimental Model, 1930–1965* (Chicago: University of Chicago Press, 2002).

10. Jessica A. Bolker, "Model Systems in Developmental Biology," *BioEssays* 17 (1995): 451–54. On the use of *Drosophila* in developmental genetics, see Evelyn Fox Keller, "*Drosophila* Embryos as Transitional Objects: The Work of Donald Poulson and Christiane Nüsslein-Volhard," *Historical Studies in the Physical and Biological Sciences* 26 (1996): 313–46.

11. Martin Rudwick, "Minerals, Strata, and Fossils," in *The Cultures of Natural His-*

tory, ed. N. Jardine, J. A. Secord, and E. C. Spary (Cambridge: Cambridge University Press, 1996), 266–86, offers a concise survey.

12.   See also Peter Galison and Alexi Assmus, "Artificial Clouds, Real Particles," in *The Uses of Experiment: Studies in the Natural Sciences*, ed. David Gooding, Trevor Pinch, and Simon Schaffer (New York: Cambridge University Press, 1989), 225–74; and Harro Maas, *William Stanley Jevons and the Making of Modern Economics* (Cambridge: Cambridge University Press, 2005), chap. 4, "Mimetic Experiments."

13.   John von Neumann, "Methods in the Physical Sciences," in *Collected Works*, ed. A. H. Taub, 6 vols. (Oxford: Pergamon, 1961–63), 6:491.

14.   Ibid.

15.   H. P. Rickman, ed., *Meaning in History: W. Dilthey's Thought on History and Society* (London: Allen and Unwin, 1961).

16.   Forrester, "If *p*, Then What? Thinking in Cases."

17.   C. C. Langdell, *A Selection of Cases on the Law of Contracts* (Cambridge: Harvard University Press, 1871), vii, quoted in Forrester, "If *p*, Then What," 15.

18.   Edward H. Levi, *An Introduction to Legal Reasoning* (Chicago: University of Chicago Press, 1949), 1.

19.   Carlo Ginzburg, "Morelli, Freud, and Sherlock Holmes: Clues and Scientific Method," in *The Sign of Three: Dupin, Holmes, Peirce*, ed. Umberto Eco and Thomas A. Sebeok (Bloomington: Indiana University Press, 1983), 97.

20.   Ibid., 92.

21.   Sir Arthur Conan Doyle, *A Study in Scarlet*, 1887, in *Sherlock Holmes: The Complete Novels and Stories*, 2 vols (New York: Bantam, 1986), 1:15.

22.   Stanislav Ulam, *Adventures of a Mathematician* (New York: Scribner's, 1976), 201.

23.   Sigmund Freud, "The Moses of Michelangelo," in *The Standard Edition of the Complete Psychological Works of Sigmund Freud*, ed. James Strachey (London: Hogarth Press, 1955), 13:222.

24.   Marcel Proust, *The Guermantes Way*, vol. 3. of *In Search of Lost Time*, ed. D. J. Enright, trans. C. K. Scott Moncrieff and Terence Kilmartin (New York: Chatto and Windus, 1992), 450.

PART I: BIOLOGY

Redesigning the Fruit Fly:
The Molecularization of *Drosophila*

MARCEL WEBER

Laboratory organisms such as the fruit fly *Drosophila melanogaster*, the soil nematode *Caenorhabditis elegans*, or the budding yeast *Saccharomyces cerevisiae* are often described as model systems or model organisms. These terms suggest that biologists cultivate and study these organisms because they provide a basis for extrapolating theoretical knowledge to other organisms, in particular *Homo sapiens*.[1] While this is clearly one of the roles that such organisms play in research,[2] this fact alone can hardly explain the widespread distribution of just a few of these organisms in laboratories all around the world. Taking an ecological perspective, we can ask what makes certain organisms so well adapted or, perhaps, adapt*able* to certain types of laboratories. Robert Kohler has shown for the case of *Drosophila* that this species entered the laboratory mostly for contingent reasons, but then turned out to be extremely well adapted and adaptable for laboratory life.[3] In addition, there now exists an impressive body of scholarship documenting how such organisms, once they have successfully colonized a few laboratories, start to affect the investigative pathways followed by the scientists.[4]

While *Drosophila* proved instrumental for the rise of genetics during the first decades of the twentieth century, the molecularization of genetics in the 1950s and 1960s largely resulted from research on microorganisms, especially *Escherichia coli* and bacteriophage. The latter organisms offered the advantage that they could be handled in the laboratory in vast numbers, thus allowing the detection of extremely rare genetic events (e.g., mutation or recombination events). Even tiny *Drosophila* was much too bulky for this task. Furthermore, most of its genes are far too complex in terms of their phenotypic effects for fine-structure mapping.[5] Even though *Drosophila* did not vanish into complete obscurity during the so-called molecular revolution, it clearly lost some of its scientific glamour. However, the fly made a spectacular comeback in the 1980s. One of the favorite experimental animals of developmental biologists today, the fruit fly even closed the race as only the second multicellular organism for which the full genomic DNA sequence became available.[6] Indeed, the fly has come a very long way since T. H. Morgan found the first mutant *white eyes* in 1910.

There are probably several factors that contributed to *Drosophila*'s come-back. Developmental biologists had kept an interest in the fly because it offered certain possibilities for the genetic analysis of development. For example, *Drosophila* can produce genetic mosaics, that is, individuals in which some lineages of somatic cells have mutated. This allowed developmental biologists to determine the fate of certain embryonic cell lineages.[7] Furthermore, the old "breeder-reactor" (Kohler's term) proved useful for a systematic screen for mutants that affect embryonic development.[8] Initially, there was little reason to believe that *Drosophila* would turn out to be a good model system for understanding the molecular basis of development in other organisms, perhaps even humans. Flies show quite a special developmental mechanism, one not only characterized by the phenomenon of metamorphosis but also by other unusual features like the syncytial stage, in which the embryo contains thousands of nuclei but no cells. *Drosophila*'s success as an experimental organism cannot be explained by its being typical or characteristic in terms of development. In fact, it is unclear in what sense *Drosophila* is really a *model* for other organisms, especially mammals.

In this article, I want to show that the main advantage responsible for *Drosophila*'s reproductive success in molecular laboratories lies in the enormous *experimental resources* associated with this organism. *Drosophila* became a powerful research tool for molecular biology because geneticists were able to mobilize these resources for molecular cloning, which gave them access to genes and gene products not confined to *Drosophila*. I also show that a *hybrid technology* was instrumental for this mobilization. The experimental resources include, first, classical genetic and cytological mapping techniques; second, highly detailed genetic maps;[9] and third, research materials such as thousands of mutants and genetically well-characterized fly strains.

In addition, I will show that the successful deployment of these experimental resources for molecular studies depended on certain theoretical concepts from classical genetics, in particular, the classical gene concept itself. I am hoping that this will shed new light on the old philosophical problem of the relationship between classical and molecular genetics.

In the following section, I will examine how the first *Drosophila* genes were isolated and characterized at the molecular level in the later 1970s and early 1980s. Then I examine the relationship between the classical genes that had already been studied by the pioneers of *Drosophila* genetics and the new molecular entities. In the section thereafter, I will examine what this case reveals about the dynamics of model system-based research. The final section of my essay draws together the main conclusions from this study of experimental practice in biology.

HOW THE FIRST *DROSOPHILA* GENES WERE CLONED

Our story begins in the midst of the cloning revolution, which originated in the Stanford University biochemistry department, where the first in vitro recombined DNA molecules were produced in 1972.[10] The scientific and technological promises, as well as ethical concerns, that the first genetically engineered organisms generated need no recounting here. In the present context, the term *cloning* designates a set of methods by which DNA fragments from any organism are inserted in vitro into a vector, that is, a small replicating unit (typically a bacterial plasmid or a bacteriophage) for amplification and subsequent molecular analysis, experimental intervention, and gene transfer.[11] It did not take long until the new so-called recombinant DNA technology—originally developed on bacteria, bacteriophage, and animal viruses—was applied to *Drosophila*, the genetically best understood multicellular organism.

In fact, some of the standard methods of gene cloning were developed using DNA isolated from *Drosophila*. For example, the laboratory of David Hogness at Stanford developed a method called "colony hybridization" that allows the isolation of any DNA fragments complementary to a given RNA molecule. Michael Grunstein and Hogness used the method to clone the genes for the *Drosophila* 18S and 28S ribosomal RNA (rRNA).[12] Recombinant plasmids containing these genes were isolated from a so-called genomic library (a set of random fragments of the entire genomic DNA of an organism inserted into a cloning vector) by hybridizing radioactively labeled RNA probes to a filter paper to which the DNA of bacterial colonies had been attached.[13]

Hogness was originally not a *Drosophila* geneticist (although he quickly became one); he was a classical molecular geneticist who had previously worked with *E. coli* and bacteriophage. However, several established *Drosophila* labs became interested in doing molecular work on the fly. One such lab was Walter Gehring's at the University of Basel, which was working on embryonic development in *Drosophila*.[14] They established a "gene bank" (the Swiss equivalent of a genomic library) by inserting DNA fragments produced by mechanical shearing of total nuclear DNA into a bacterial plasmid.[15] Then they used the colony hybridization technique on this gene bank to isolate the *Drosophila* 5S rRNA genes.[16] Gehring's laboratory later turned to the heat shock genes. Like most cells, *Drosophila* expresses a number of specific proteins after a heat shock and shuts down the expression of most other genes. Thus mRNA isolated from heat-shocked cells, together with the colony hybridization technique, provided an easy way to isolate the genes encoding the heat shock proteins.[17]

The colony hybridization method requires a purified RNA molecule complementary to the gene to be cloned. In the case of the rRNA and heat shock genes, abundant RNA species were available for this task. However, not all genes that interested the geneticists offered such ready access. A much more difficult cloning task involved the *white* gene—the oldest *Drosophila* gene known, as the first *white* mutation was described and localized to the X-chromosome by T. H. Morgan in 1910. This gene, characterized by mutations with a variety of phenotypic effects, had eluded all attempts to determine its physiological function. Therefore Gerald Rubin's laboratory at the Carnegie Institution used a trick called "transposon tagging" to clone the *white* locus. They made use of a particular *white* allele (*white-apricot* or *w^a*), which had been shown to result from an insertion of the transposable element *copia* into the *white* locus.[18] Since *copia* DNA was easy to isolate (it produces an abundant mRNA species that contaminated many cloning experiments and was thus accidentally isolated as a false positive in many colony hybridization screens), Rubin's laboratory was able to use *copia* as a molecular probe to isolate cloned DNA fragments containing sequences from the *white* locus.[19]

Gehring's lab used a different strategy to clone the *white* locus. They made use of the existence of a *Drosophila* strain in which the *white* locus had moved with a large transposon[20] from the X-chromosome to a new location on chromosome 3, which happened to be in the vicinity of the already cloned heat shock genes.[21] They then used a technique called "chromosomal walking" (see below) to isolate sequences from the *white* locus. This was possible because they had already cloned the heat shock genes, which served as a starting point for their chromosomal walk. In order to confirm the identity of the cloned DNA, they showed that it hybridized in situ to the known cytological location of the *white* locus. This technique of in situ hybridization proves important for our story and I therefore briefly describe it here.

The salivary glands of *Drosophila* larvae contain giant chromosomes composed of thousands of copies of nuclear DNA. These so-called polytene chromosomes have been used for cytological mapping since the 1930s. Cytological maps based on the specific banding patterns of polytene chromosomes were shown to be colinear with linkage maps in the 1930s.[22] The method of in situ hybridization allows the localization of specific DNA fragments on polytene chromosome preparations. In this method, the DNA fragment to be localized is radioactively labeled and subsequently hybridized to polytene chromosomes. The chromosomal locations in which the DNA probe hybridized can then be visualized using autoradiography.

Direct evidence that these cloned DNA fragments really contain the *white* gene was only obtained two years later: Using a transposable element (P-element) as a vector for germ-line transformation that was developed in Gerald Rubin's laboratory,[23] Gehring and his coworkers demonstrated that the putative *white*-locus sequences are capable of rescuing the *white-minus* phenotype, that is, *white*-mutants transformed with the cloned DNA showed the red eye color of the wild type.

The cloning strategies used for the rRNA and heat shock genes relied on the availability of the gene products of these genes (rRNA and heat shock protein mRNA, respectively). The Rubin lab's strategy to clone the *white* gene made use of the fact that one of the *white* alleles was the result of a transposable element insertion, so that cloned DNA containing this transposable element could be used to "fish" for DNA fragments containing the *white* gene. Since these are rather serendipitous circumstances that may not obtain in the case of other genes of interest, Hogness's laboratory developed an ingenious method to clone *Drosophila* DNA sequences about which nothing is known save their chromosomal location. This method came to be known as the chromosomal walking mentioned earlier.

Chromosomal walking makes full use of the powerful resources of *Drosophila* cytogenetics. A chromosomal walk can start with any fragment of previously cloned *Drosophila* DNA located not too far from the region to be cloned. In a first step, the cytological location of the cloned DNA sequence is determined by in situ hybridization to polytene chromosomes. In the following step, the starting DNA fragment is used to screen a genomic library for random fragments overlapping the starting fragment. The new fragments are mapped by in situ hybridization. Then they can be aligned to the starting fragment by restriction endonuclease mapping in order to determine the fragment farthest from the starting point in the direction of the walk.[24] This second step is repeated many times and generates overlapping cloned DNA fragments that lie farther and farther from the starting point. The end point of a chromosomal walk is the region of interest, that is, the known or assumed location of a gene or gene complex of interest (as determined by classical cytogenetic mapping). In order to save time, it is possible (step 3) to "jump" large distances on the chromosome by using chromosomal rearrangements such as inversions. Rearrangements such as small deletions can also be used to narrow down the position of any cloned DNA fragment. For instance, if such a fragment fails to hybridize to a polytene chromosome carrying a deletion, this indicates that the cloned sequence lies within the break points of the deletion.

To my knowledge, the technique of using in situ hybridization in order to clone DNA sequences about which only their cytological location is known was unique to *Drosophila*. Hogness's laboratory first used it to clone the *rosy* and *Ace* genes.[25] As Welcome Bender, Pierre Spierer, and Hogness report in their paper, while their "walk" was in progress, they learned from E. B. Lewis of inversions with end points in the Bithorax complex and in the region in which they were walking and used these to jump into the Bithorax region.[26] Thus, rather accidentally, the first homeotic gene complex was cloned.[27]

One of the next genes to be cloned by chromosomal walking was *Antennapedia* (*Antp*), another homeotic gene responsible for bizarre mutations, such as flies with legs on their heads. Two (competing) laboratories were involved in this strenuous effort, which, in the case of the Gehring group, took more than three years to complete.[28] They cloned the *Antp* gene by a long chromosomal walk that made use of the chromosomal inversion *Humeral*, used to jump from a previously cloned region into the *Antp* region.[29] It is the availability of mutants such as *Humeral*, which had previously been characterized by classical genetic methods, that made *Drosophila* such a powerful system for molecular research.

The laboratory of Thomas Kaufman at Indiana University used a somewhat different strategy, which also involved much chromosomal walking, to isolate clones spanning the *Antp* region.[30] Both laboratories were able to use their chromosomal walk in the *Antp* region to clone another gene located in close vicinity (in fact, it is part of the Antennapedia gene complex[31]): *fushi tarazu* or *ftz*. The Basel and Indiana groups independently found a sequence homology between *Antp*, *ftz*, and *Ultrabithorax* (one of the genes of the Bithorax gene complex, which had also been cloned). The homologous sequence turned out to be a highly conserved sequence element of 180 base pairs length. It was named the "homeobox" and had a great impact on molecular studies of development. Using cloned *Drosophila* sequences containing the homeobox as hybridization probes, homeobox genes from a great range of organisms including humans and plants have been isolated in a very short time.[32]

I shall now turn to the question of what exactly has been "cloned" in these cases.

THE CLONING OF WHAT?

So far, I have taken the notion of cloning genes to be unproblematic. I now want to suggest that it raises a conceptual puzzle. The concept of the gene has been the subject of much philosophical debate.[33] There is a consensus that the meaning of the term *gene* is historically highly variable and that it continues to resist attempts to give it a straightforward definition (which is not unusual for key concepts in science). Disagreement exists on the relationship between the so-called classical gene concept of the Morgan school, the "neoclassical" concept that arose after DNA was recognized as the genetic material, and the molecular gene concept. While some think that this relationship is, by and large, one of conceptual refinement with referential stability, others hold the classical and molecular concepts to be incommensurable. I suggest that my account of the cloning of the first *Drosophila* genes adds a new twist to this problem.

While previous discussions of the classical and molecular gene concepts mostly remained at the abstract level of conceptual analysis, our *Drosophila* story allows us to follow the trajectories of particular genes from the early days of classical genetics through different periods of *Drosophila* genetics. In particular, we can follow the fate of some genes into the molecular period. As I have shown, drosophilists have cloned a number of genes already known to Morgan and his students, for example, *white*, *rudimentary*, Bithorax, or *achaete-scute*. However, on the background of the recent debates on the gene concept, the notion that, for example, the *white* gene was cloned raises a conceptual puzzle, for *white* is a classical gene, and what Rubin, Gehring, and their coworkers have isolated seems to be a molecular entity, namely, a piece of DNA. What is it *exactly* that they cloned?

In order to answer this question, I begin by examining how classical genes are identified. In classical genetics, the existence of genes is inferred solely from the existence of *mutations*, which were recognized by their phenotypic effects. A widely used criterion for deciding whether two mutations affect the same or two different genes was *complementation*, that is, reversion of the mutant phenotype to wild type when the two mutations are present in *trans* (i.e., on two different but homologous chromosomes). What complementation shows is the *functional sufficiency* of a combination of alleles to provide the full function of a gene.[34] Initially, it was thought that the occurrence of recombination between two genetic markers can also be used as a criterion for delimiting genes. The recognition that recombination can occur within genes rendered this criterion useless and left complementation as the main criterion to be used to identify genes.[35]

However, there remained difficulties with this criterion, difficulties that arose in particular with so-called complex loci. Such loci are characterized by a plethora of different phenotypes associated with different alleles and by complex interactions between these alleles. Heterozygotes of two mutant alleles at complex loci frequently show intermediate or completely different phenotypes in *trans* (they are also referred to as "pseudoalleles" for this reason). This kind of *interallelic* or *intragenic* complementation makes it difficult to decide how many genes are actually affected in such a series of mutations. In some cases, it was impossible to discern any systematic relationship between the genetic interactions of different mutations and their map location. In contrast to the linear complementation maps found by Seymour Benzer in bacteriophage T4, some complementation data from complex loci were more consistent with circular maps. However, these maps were unlikely to have any representational content with respect to gene structure.[36] Nevertheless, Petter Portin argues that there exist fairly robust criteria for distinguishing intragenic from intergenic complementation and that, therefore, the complementation criterion for identifying genes is sufficient in most cases.[37]

In fact, most of the loci discussed here can be viewed as complex loci, even those that are not traditionally classified in this manner.[38] The question I would like to examine now is how the molecular analysis of these loci—which became possible once they had been cloned—is related to their premolecular, genetic analysis. Let us look at a few examples in detail.

White

As we have seen, DNA sequences from the *Drosophila white* locus were isolated (1) by "transposon tagging" using an allele containing an insertion of the transposable element *copia*, which had been cloned previously; and (2) by using a transposition mutant (which had been mapped by classical cytogenetic methods) in which the *white* locus had moved into the vicinity of the previously cloned heat shock genes. Furthermore, transformation experiments with the cloned DNA into *white-minus* flies and the resulting rescue of the *white* phenotype showed that the cloned fragments contained all the sequences for normal gene function. This may be viewed as some sort of a complementation test; it demonstrates functional sufficiency. Some cloned DNA fragments were also shown to contain the chromosomal region bracketing the *white* locus according to cytogenetic maps. A full DNA sequence analysis of the *white* locus revealed a 2.6 kb transcription unit with five exons encoding a hydrophobic (probably membrane-bound) protein.[39] In addition, it was possible to provide a molecular analysis of some of the known mutant alleles of *white*. For instance, a group of *white-minus* alleles that clustered at one end of the locus and that did not appear to obliterate wild-type function completely had been hypothesized to be so-called regulatory mutations by classical geneticists. Indeed, molecular analysis showed these mutations to reside in the upstream region of *white*. To give another example, the molecular analysis of *white-ivory* confirmed a result already obtained on the basis of linkage mapping, namely, that this mutation constitutes an intragenic tandem duplication. Remarkably, it was shown that some alleles contained insertions into the gene's introns, suggesting that alterations within an intron may affect gene expression.

Thus there appears to be a fairly good correspondence between the molecular analysis and previous genetic studies of the *white* locus. What the researchers initially cloned was a chromosomal fragment identified as the *white* locus using classical cytogenetic methods. On the basis of complementation analysis, this locus was assumed to contain one gene, the boundaries of which had been mapped cytogenetically. Molecular analysis then showed that this chromosomal fragment indeed contained a generic molecular gene,[40] that is, a protein-coding DNA sequence in five exons and an upstream regulatory region. Thus, in the case of *white*, the classical and the molecular gene concepts pick out approximately the same chromosomal region—assuming that the regulatory sequences are considered to be part of a molecular gene. If all genes were like this, those stressing referential stability of the gene concept (in spite of meaning variance) would be right. Unfortunately, things are not always as simple. This will become evident when we turn to the complex loci.

Rudimentary

This classic complex locus was also first described by Morgan in 1910. The alleles at this locus exhibit a complex complementation pattern, with some alleles exhibiting intragenic complementation and others showing no complementation at all. The locus encodes the enzymatic activities for the first three steps in pyrimidine biosynthesis. It had been shown that all three activities are encoded by a single gene in *Drosophila* and in hamsters, while up to three different genes are needed in other organisms. Different genetic studies of this locus have found different numbers of complementation groups, that is, classes of alleles that behave identically in a complementation test.[41] It was expected, from the trifunctional nature of the enzyme encoded, that three complementation units (i.e., inferred functional parts of the gene) would be found, with each enzymatic activity corresponding to a complementation unit. However, at least seven units were found by complementation analysis. This was attributed to protein-protein interactions between the subunits in the multimeric protein because the complementation observed appeared to be intragenic.

For our purposes, it is interesting to note that classical genetic analysis indicated the presence of a *single* gene (in *Drosophila*), in spite of the complex behavior of the locus in test crosses of different alleles. When the locus was cloned (using a previously available probe of hamster DNA), a single transcription unit was indeed found.[42] Thus we seem to have here another example in which classical genetic methods correctly predicted a protein-coding molecular gene.

Bithorax

This locus was studied intensely before molecular cloning methods became available in *Drosophila*, most famously by Edward B. Lewis (this work earned him a Nobel Prize). The genetics of this locus turned out to be exhilaratingly complex, especially because of the existence of several mutations that enhance or silence the effect of other mutations at the same locus. It was thus assumed that the locus contains a whole gene complex rather than a single gene. Lewis, on the basis of his famous theoretical model of the determination of segment identity, postulated "at least eight genes."[43] However, an alternative model of the genetic organization of the Bithorax complex was provided by Ginés Morata's laboratory.[44] They isolated Bithorax mutants from a mutagenesis experiment and crossed these with strains carrying large Bithorax deletions. Crosses that failed to produce viable larvae were used to identify lethal alleles. These lethal alleles were then examined for complementation. Three complementation groups were found and named *Ultrabithorax* (*Ubx*), *abdominal-A* (*abd-A*), and *Abdominal-B* (*Abd-B*).

This simplified model of the Bithorax locus was not immediately accepted.[45] However, the molecular analysis eventually revealed three protein-coding genes corresponding to *Ubx*, *abd-A*, and *Abd-B*. Thus the classical complementation criterion for genes successfully predicted the molecular protein-coding genes, while Lewis's analysis, based solely on the phenotypic effects of different Bithorax mutants, predicted too many genes.

Antennapedia

Like Bithorax, this homeotic locus was assumed to contain a gene complex. However, in contrast to Bithorax, a single classical gene named *Antennapedia* had also been identified on the basis of genetic analysis. In fact, the molecular analysis found a single protein-coding gene (but two nested transcription units) located in the genetic map region bracketed between known inversion break points defined by different mutant *Antp* alleles. The Antennapedia complex contains a number of additional genes, for example, the segmentation gene *fushi tarazu* (*ftz*), which mapped between the *Antp* and the *Deformed* locus. The *ftz* gene was also cloned in both Gehring's and Kaufman's laboratories,[46] and was shown to contain a single protein-coding gene. Thus in the case of the Antennapedia complex, molecular studies confirmed most of the expectations from classical genetic analysis.

Achaete-scute

This was one of the first loci that revealed the full extent of the possible genetic complexity of *Drosophila*, as was shown by H. J. Muller, N. P. Dubinin, and others in the 1930s. It is a classical example of a step-allelic series,[47] and first-generation *Drosophila* geneticists had hoped to learn something about the internal structure of genes by complementation analysis of the *achaete-scute* series of mutations. However, these efforts were met with little success.[48] More recently, mapping of various *achaete-scute* lesions led to the subdivision of the locus into four regions: *achaete, scute α, lethal of scute,* and *scute β*.[49] A fifth region, named *scute γ*, was postulated by Christine Dambly-Chaudière and Alain Ghysen.[50] The locus was cloned in 1985 and revealed a complex arrangement of nine transcription units, the relationship of which to the genes was not clear at first.[51] However, it was eventually possible to identify one transcription unit for each of the *achaete, scute α,* and *lethal of scute* regions.[52] A fourth transcript was located to the opposite end of the region and named *asense*. The *scute β*, separated from *scute α* by *lethal of scute*, is now thought to be a *cis*-regulatory element affecting the expression of the *scute* protein. Finally, a full DNA sequence analysis of the cloned *achaete-scute* locus revealed four homologous protein-coding genes (all encoding transcription factors), corresponding to *achaete, scute, lethal of scute,* and *asense*. It is thought today that the complete function ascribed to the *achaete-scute* region on the basis of genetic experiments is contained in these four transcription-factor genes.[53] The complexity of the locus stems from the interspersion of coding sequences and regulatory sites, such that different regulatory sites act on the same gene and the regulatory sites act on different genes. This example also shows the complex nature of eukaryotic genes, which comprise not just protein-coding regions but often include regulatory sequences involving a far larger segment of the chromosome.

Thus, although this locus shows the limitations of premolecular mapping techniques most acutely, classical genetic analysis came close to predicting the molecular protein-coding genes of the *achaete-scute* locus;[54] only *scute β* turned out to be a *cis*-acting regulatory element instead of a full-blown gene.

In the light of these examples, we can now address the question posed, namely, what it was that the *Drosophila* molecular geneticists cloned. My analysis suggests a perhaps rather unexpected result: in many cases, what was cloned initially were classical genes, that is, chromosomal fragments that had been identified by a combination of genetic and cytological mapping, as well as by a complementation analysis of mutation series. The molecular *Drosophila* geneticists then used different methods to make sure that the DNA fragments they had cloned contained all the sequences necessary for the function of the gene in question, including upstream and downstream regulatory regions. The severest test for whether a cloned DNA fragment contains all the sequences necessary for gene function is a germ-line transformation experiment of mutant flies with this cloned DNA fragment (such an experiment was performed, e.g., with the cloned *white* fragments or with *fushi tarazu*). If the transformation "rescues" the mutation, the DNA inserted must contain a fully functional gene with all the necessary regulatory regions. This is a *functional* test, a molecular version of the classical complementation test. Thus molecular *Drosophila* geneticists basically used the classical criteria for identifying genes in order to locate the genes sought on the DNA molecule.

Once the identity of the cloned DNA sequences had been confirmed, the geneticists analyzed them for *molecular* genes, that is, they looked for protein-coding exons, as well as promoters, upstream and downstream regulatory regions, and so on. In most cases, the chromosomal regions expected, on the basis of classical genetic tests, to contain a single gene were indeed shown to contain a single molecular protein-coding gene (including regulatory regions). Even the genes residing at the complex loci had been correctly identified by complementation analysis (with some minor exceptions such as *scute*; see above), thus confirming Portin's point that the complementation criterion is sufficient for identifying genes.

But the main point I wish to emphasize is that without the classical gene concept and the associated mapping methods, most of the molecular genes would have been much more difficult to find in genomic libraries from *Drosophila*, which contain millions of different DNA fragments. Somewhat paradoxically perhaps, it was the *classical* gene concept that directed geneticists to the *molecular* genes of *Drosophila*. Thus I am suggesting that the classical gene concept and its associated operational criteria, although they had long been replaced by the molecular gene concept at the *theoretical* level,[55] still played a role in experimental practice.

DROSOPHILA AS A MOLECULAR RESEARCH TOOL

The story of the cloning of the first *Drosophila* genes allows us to address a number of issues that have to do with the role of experimental systems and model organisms in biological research. The main reason why *Drosophila* became such a powerful research tool for molecular biology now seems fairly clear: it is, first and foremost, the availability of well-defined laboratory strains with thousands of known mutations that had been mapped by linkage and cytological mapping. As Kohler has shown,[56] *Drosophila* became, as it were, ecologically adapted for life in genetics laboratories around the world because of its ability to function as a "breeder-reactor" for mutants that could be systematically classified by the linear arrangement of mutations on increasingly detailed chromosomal maps. My account suggests that the momentum that *Drosophila* had acquired as an experimental system basically carried over into the molecular era. The question is, then, how this was possible. How could the enormous experimental resources of the *Drosophila* system be mobilized for molecular research? Before I turn to this issue, I shall briefly discuss some of the advantages that *Drosophila* offered in comparison to other model organisms such as *C. elegans*, the mouse, or *Xenopus*.

An important factor was the possibility to clone genes about which nothing was known except their chromosomal location ("positional cloning"). Specifically, nothing was known about the gene *products*. For example, this was the case for the *white* locus, for the homeotic loci of the Bithorax and Antennapedia complexes, and for some genes involved in neurogenesis such as *Notch* or *achaete-scute*. Although the specific phenotypic effects of mutations at these loci gave some clues as to their role in development, absolutely nothing was known about the proteins encoded by these genes. Furthermore, these gene products and their mRNAs are present at extremely low concentrations, which made their biochemical isolation difficult, if not impossible.[57] What made the cloning of these genes possible is the fact that their chromosomal positions had been determined to a great degree of precision by cytogenetic mapping.[58]

A key technique was Hogness's method of chromosomal walking, which allowed geneticists to isolate DNA fragments from a genomic library using previously cloned DNA fragments. The method of in situ hybridization in polytene chromosomes allowed geneticists to navigate on the chromosome by localizing newly cloned DNA fragments on the polytene chromosomes. The large number of available and cytologically mapped inversion, deletion, and insertion mutants could be used to speed up the process of chromosomal walking (by jumping large distances) and to pinpoint the location of the DNA to be cloned much more precisely. No other model organism offered these powerful classical extensions of the standard molecular cloning repertoire. Another important technique, developed in Rubin's laboratory, was the possibility of germ-line transformation of *Drosophila* using the transposable P-element. This made it possible to reintroduce cloned DNA fragments in order to confirm their functional sufficiency or in order to alter the expression of specific genes.

Thus, in part, it was the enormous experimental resources of premolecular *Drosophila* genetics that gave the fruit fly a head start over other experimental organisms like *C. elegans* or the mouse. The chromosomal walking method of positional cloning provided an entry point to some genes and their products that, in all likelihood, could not have been isolated using other experimental organisms. The best examples are the homeobox genes. Although a few homeotic mutants were known in other organisms, these organisms lacked the kind of experimental resources that the *Drosophila* system offered. But once the first homeotic genes had been cloned in *Drosophila*, researchers were able to use them as hybridization probes to screen for homologous genes in other organisms by using so-called zooblots.[59]

I propose that the molecular transition in *Drosophila* genetics succeeded so remarkably because of a *hybrid experimental system*.[60] Such hybrid systems combine the practices and research materials derived from some parent technologies and, at the same time, transform them into something new. The hybrid system that, as I suggest, played an important role in the rise of *Drosophila* as a molecular research tool has two parents: first, classical *Drosophila* genetics and, second, recombinant DNA technology. Classical *Drosophila* genetics had been developed by several generations of geneticists who had isolated and characterized thousands of mutant strains, thus producing ever more detailed chromosomal maps showing the location of the mutations. Cytological mapping of polytene chromosomes proved to be an especially powerful technique. This technique used the unique patterns of bands at specific locations on polytene chromosomes, which were correlated with mutant phenotypes in order to determine the location of genetic loci, as well as of chromosomal rearrangements such as deletions, inversions, and translocations.[61] Recombinant DNA technology, by contrast, was mainly developed by work using *E. coli*, bacteriophage, and animal viruses as experimental systems. The basic tools of this technology are a set of purified enzymes that modify DNA in various ways, for example, restriction endonucleases recognizing specific sequences and hydrolyzing DNA phosphodiester bonds at specific locations, or DNA ligase that can be used to join DNA fragments in vitro.

I suggest that these two parent technologies—classical *Drosophila* genetics and recombinant DNA technology—were used to create a hybrid system for the identification and isolation of genes. An important part of the hybrid system was the method of in situ hybridization of radioactively labeled DNA to polytene chromosomes, which was originally developed on chromatin from *Xenopus* oocytes.[62] This technique greatly expanded the capacities of the cytological mapping system by allowing researchers to visualize the position of specific DNA sequences on polytene chromosomes. Thus one of the most powerful mapping approaches available in *Drosophila* was adapted to the needs of molecular cloning. The technique of chromosomal walking, which allowed the isolation of DNA from specific locations on the chromosome, relied heavily on in situ hybridization. In addition, geneticists made clever use of available inversion and deletion mutants (which had been mapped cytologically) in order to jump over large chromosomal regions and in order to determine the exact position of their DNA clones between the break points of known chromosomal rearrangements. This demonstrates how the hybrid system was able to integrate the practices as well as the research materials from both the classical and molecular research traditions in genetics. It enabled the scientists to mobilize the experimental resources that came with these traditions, for example, mutants, strains, maps, DNA-modifying enzymes, preparation techniques, and so on to create a powerful new experimental system.[63]

Any well-defined DNA fragment cloned constituted a potential resource for further experimental work because it could serve as a tool for the cloning of other DNA fragments. For example, an already cloned DNA fragment could serve as a starting point of chromosomal walks to isolate sequences located in a near chromosomal region. This way, cloned DNA fragments were used to clone other, totally unrelated sequences. In order to isolate homologous sequences, cloned DNA was used as a hybridization probe in colony hybridization screens. This method could be applied to DNA isolated from any organism, a fact that proved crucial for isolating a host of genes from other organisms (including humans) using cloned DNA fragments from *Drosophila*. Thus the more genes had been cloned, the easier it became to clone genes. With each new gene cloned, the *Drosophila* system was transformed from a breeder-reactor for isolating and mapping mutants into a molecular research tool for the production of specific recombinant DNA molecules.

CONCLUSIONS

Beginning in the 1970s, molecular cloning techniques transformed the gene from an abstract entity characterized by the phenotypic effects of mutations (or, later on, by its role in RNA- and protein metabolism) into an object of direct experimental intervention. With these techniques, it became possible to isolate genes, analyze them at the level of DNA sequences, transfer them into other, totally unrelated organisms, and to introduce specific alterations. Together with yeast (about which a similar story could be told), *Drosophila* emerged as one of the major eukaryotic experimental organisms in this transition. Part of the explanation for *Drosophila*'s rapid and extremely successful molecular transition lies in the fact that, using Hans-Jörg Rheinberger's terms,[64] a hybrid experimental system allowed researchers to appropriate the enormous experimental resources of classical *Drosophila* genetics. By experimental resources I mean not just the knowledge of mapping techniques and the like but also *material* resources such as the thousands of genetically defined mutant strains. The hybrid system allowed researchers to mobilize these resources for cloning experiments, and thus to transform them into molecular research tools. In the process, *Drosophila* was transformed from a system for the production of mutants and genetic maps into a system for the production of specific recombinant DNA molecules.

My account of the cloning of the first *Drosophila* genes also reveals that at least for some of the genes subjected to extensive classical genetic analysis, the classical gene *concept* played a crucial role in isolating the specific DNA fragments containing these genes. Whatever the implications for the semantic relationship between the classical and molecular gene concepts may be, it is clear that it was the classical gene concept and its operational criteria that led to the identification of a number of molecular, protein-coding genes. The classical *gene* may have ceased to exist, but the classical gene *concept* and the associated experimental techniques and operational criteria proved instrumental for the identification of molecular genes and, therefore, for many of the recent advances made in understanding genetic processes at the molecular level.[65] Perhaps to a surprisingly large extent, the initial research agenda of molecular studies in *Drosophila* was set by the strong experimental tradition (beginning around 1910) that had used this organism for classical genetic analysis. Theoretical concepts, experimental techniques, and research materials are closely intertwined in this tradition.

NOTES

I wish to thank Walter Gehring for the interview and Jay Aronson, Paul Hoyningen-Huene, and Daniel Sirtes for helpful comments. This article also benefited from discussions with audiences at the University of Minnesota (Ken Waters's graduate student seminar in the philosophy of biology, in the fall of 2000) and at the March 2001 "Mapping Cultures" workshop at the Max Planck Institute for the History of Science in Berlin.

1. Kenneth F. Schaffner, "Extrapolation from Animal Models: Social Life, Sex, and Super Models," in *Theory and Method in the Neurosciences,* ed. Peter K. Machamer, Rick Grush, and Peter McLaughlin (Pittsburgh: University of Pittsburgh Press, 2001), 200–230.

2. Marcel Weber, *Philosophy of Experimental Biology* (Cambridge: Cambridge University Press, 2005), chap. 6; Marcel Weber, "Under the Lamppost: Commentary on Schaffner," in Machamer, Grush, and McLaughlin, *Theory and Method in the Neurosciences,* 231–49.

3. Robert E. Kohler, *Lords of the Fly: Drosophila Genetics and the Experimental Life* (Chicago: University of Chicago Press, 1994).

4. Adele E. Clarke and Joan H. Fujimura, eds., *The Right Tools for the Job: At Work in Twentieth-Century Life Sciences* (Princeton: Princeton University Press, 1992); Richard M. Burian, "How the Choice of Experimental Organism Matters: Biological Practices and Discipline Boundaries," *Synthese* 92 (1992): 151–66; Richard M. Burian, "How the Choice of Experimental Organism Matters: Epistemological Reflections on an Aspect of Biological Practice," *Journal of the History of Biology* 26 (1993): 351–67; Richard M. Burian, "'The Tools of the Discipline: Biochemists and Molecular Biologists'; A Comment," *Journal of the History of Biology* 29 (1996): 451–62; Robert E. Kohler, "Systems of Production: *Drosophila, Neurospora* and Biochemical Genetics," *Historical Studies in the Physical and Biological Sciences* 22 (1991): 87–130; Robert E. Kohler, "*Drosophila*: A Life in the Laboratory," *Journal of the History of Biology* 26 (1993): 281–310; Bonnie Tocher Clause, "The Wistar Rat as a Right Choice: Establishing Mammalian Standards and the Ideal of a Standardized Mammal," *Journal of the History of Biology* 26 (1993): 329–49; Gerald L. Geison and Angela N. H. Creager, "Introduction: Research Materials and Model Organisms in the Biological and Biomedical Sciences," *Studies in History and Philosophy of Biological and Biomedical Sciences* 30 (1999): 315–18; Angela N. H. Creager, *The Life of a Virus: Tobacco Mosaic Virus as an Experimental Model, 1930-1965* (Chicago: University of Chicago Press, 2002); Frederic L. Holmes, "The Old Martyr of Science: The Frog in Experimental Physiology," *Journal of the History of Biology* 26 (1993): 311–28; Muriel Lederman and Richard M. Burian, "Introduction: The Right Organism for the Job," *Journal of the History of Biology* 26 (1993): 235–37; Muriel Lederman and Sue A. Tolin, "Ovatoomb: Other Viruses and the Origins of Molecular Biology," *Journal of the History of Biology* 26 (1993):

239–54; Karen A. Rader, "Of Mice, Medicine, and Genetics: C. C. Little's Creation of the Inbred Laboratory Mouse," *Studies in History and Philosophy of Biological and Biomedical Science* 30 (1999): 319–43; Hans-Jörg Rheinberger, "*Ephestia*: The Experimental Design of Alfred Kühn's Physiological Developmental Genetics," *Journal of the History of Biology* 33 (2000): 535–76; William C. Summers, "How Bacteriophage Came to Be Used by the Phage Group," *Journal of the History of Biology* 26 (1993): 255–67; Doris T. Zallen, "The 'Light' Organism for the Job: Green Algae and Photosynthesis Research," *Journal of the History of Biology* 26 (1993): 269–79.

5. Marcel Weber, "Representing Genes: Classical Mapping Techniques and the Growth of Genetic Knowledge," *Studies in History and Philosophy of Biological and Biomedical Sciences* 29 (1998): 295–315.

6. Mark D. Adams et al., "The Genome Sequence of *Drosophila melanogaster*," *Science* 287 (2000): 2185–95.

7. Adam S. Wilkins, *Genetic Analysis of Animal Development* (New York: Wiley, 1993).

8. Christiane Nüsslein-Volhard and Eric Wieschaus, "Mutations Affecting Segment Number and Polarity in *Drosophila*," *Nature* 287 (1980): 795–801.

9. In another work, I argue that the role of genetic maps changed from a *representational* to a *preparative* mode in the course of developing *Drosophila* into a molecular research tool; see Marcel Weber, "Walking on the Chromosome: *Drosophila* and the Molecularization of Development," in *From Molecular Genetics to Genomics: The Mapping Cultures of Twentieth-Century Genetics*, ed. Jean-Paul Gaudillière and Hans-Jörg Rheinberger (London: Routledge, 2004), 63–78.

10. Michel Morange, *A History of Molecular Biology* (Cambridge: Harvard University Press, 1998), chap. 16.

11. The term *molecular cloning* is also used. It has the advantage of avoiding confusion with Dolly, the sheep.

12. Michael Grunstein and David S. Hogness, "Colony Hybridization: A Method for the Isolation of Cloned DNAs That Contain a Specific Gene," *Proceedings of the National Academy of Sciences of the United States of America* 72 (1975): 3961–65.

13. The term *hybridization* refers to the experimental production of double-stranded nucleic acid molecules in which the two strands are derived from different sources. The technique makes use of the strong disposition of complementary nucleic acid molecules (which is also crucial for their "informational" properties) to anneal to a double helix. Under certain conditions, it is possible to hybridize single-stranded RNA to a complementary DNA, which forms the basis of colony hybridization and many other molecular techniques.

14. Gehring contracted the cloning know-how in the form of a postdoc, Paul Schedl, who had been trained at Stanford. See Walter J. Gehring, *Master Control Genes in Development and Evolution: The Homeobox Story* (New Haven: Yale University Press, 1998), 41.

15. Walter J. Gehring, "Establishment of a *Drosophila* Gene Bank in Bacterial Plas-

mids: Elizabeth Goldschmidt Memorial Lecture," *Israel Journal of Medical Sciences* 14 (1978): 295–304.

16. Spyros Artavanis-Tsakonas et al., "The 5S Genes of *Drosophila melanogaster*," *Cell* 12 (1977): 1057–67.

17. Paul Schedl et al., "Two Hybrid Plasmids with *D. melanogaster* DNA Sequences Complementary to mRNA Coding for the Major Heat Shock Protein," *Cell* 14 (1978): 921–29; Spyros Artavanis-Tsakonas et al., "Genes for the Seventy Thousand Dalton Heat Shock Protein in Two Cloned *D. melanogaster* DNA Segments," *Cell* 17 (1979): 9–18; Laurie Moran et al., "Physical Map of Two *D. melanogaster* DNA Segments Containing Sequences Coding for the Seventy Thousand Dalton Heat Shock Protein," *Cell* 17 (1979): 1–8.

18. Walter J. Gehring and Renato Paro, "Isolation of a Hybrid Plasmid with Homologous Sequences to a Transposing Element of *Drosophila melanogaster*," *Cell* 19 (1980): 897–904.

19. Paul M. Bingham, Richard Levis, and Gerald M. Rubin, "Cloning of DNA Sequences from the *White* Locus of *D. melanogaster* by a Novel and General Method," *Cell* 25 (1981): 693–704.

20. This transposon, which contained the *roughest* locus in addition to *white*, was found by classical linkage mapping; see G. Ising and C. Ramel, "The Behavior of a Transposing Element in *Drosophila melanogaster*," in *The Genetics and Biology of Drosophila*, ed. M. Ashburner and E. Novitski (New York: Academic Press, 1976), 947–54. More than one hundred transposition events involving this transposon were detected.

21. Michael L. Goldberg, Renato Paro, and Walter J. Gehring, "Molecular Cloning of the *White* Locus Region of *Drosophila melanogaster*," EMBO *Journal* 1 (1982): 93–98; Renato Paro, "Molecular Cloning of Large Transposable Elements Carrying the *White* Gene of Drosophila Melanogaster" (PhD diss., University of Basel, 1982).

22. Weber, "Representing Genes," 296.

23. Gerald M. Rubin and Alan C. Spradling, "Genetic Transformation of *Drosophila* with Transposable Element Vectors," *Science* 218 (1982): 348–53.

24. Endonuclease mapping constitutes another important technique that uses the unique distribution of cleavage sites for different restriction enzymes in order to physically map DNA molecules. The DNA fragment to be mapped is cut with a number of restriction enzymes, and the resulting fragments are separated by molecular weight by gel electrophoresis. Since the molecular weight of the fragments is proportional to their length, the specific pattern of restriction fragments can be used to construct a map of the restriction sites on the DNA fragment.

25. Actually, the products of these genes were known from classical studies in biochemical genetics (xanthine dehydrogenase and acetylcholinesterase, respectively). However, this knowledge could not then be used for cloning because the concentrations of the corresponding mRNAs were too low to permit isolation of specific RNAs for colony hybridization. By contrast, some other so-called house-

hold genes like *Adh* (alcohol dehydrogenase) were cloned by prior identification of an mRNA; see Cheeptip Benyajati et al., "Alcohol Dehydrogenase in *Drosophila*: Isolation and Characterization of Messenger RNA and cDNA Clone," *Nucleic Acids Research* 8 (1980): 5649–67.

26. Welcome Bender, Pierre Spierer, and David S. Hogness, "Chromosomal Walking and Jumping to Isolate DNA from the *Ace* and *Rosy* Loci and the Bithorax Complex in *Drosophila melanogaster*," *Journal of Molecular Biology* 168 (1983): 17–33.

27. *Homeotic* is a term introduced by William Bateson in order to refer to mutants in which some body part of an animal is transformed into a structure that belongs to a more anterior or more posterior location. Mutations at the *Drosophila* Bithorax locus affect the thoracic segments of the fly body. Some mutants even have a second pair of wings—a most embarrassing condition for a member of the order Insecta Diptera, which means "two-winged insects." Because of their drastic phenotypic effects, homeotic genes were always thought to play a crucial role in embryonic development.

28. Gehring, *Master Control Genes in Development and Evolution*, 46.

29. Richard L. Garber, Atsushi Kuroiwa, and Walker J. Gehring, "Genomic and cDNA Clones of the Homeotic Locus Antennapedia in *Drosophila*," EMBO *Journal* 2 (1983): 2027–36.

30. Matthew P. Scott et al., "The Molecular Organization of the Antennapedia Locus of *Drosophila*," *Cell* 35 (1983): 763–76.

31. Note that the *Antp* gene itself is written in italics, while the Antennapedia gene complex is written in roman letters.

32. Walter J. Gehring, "Homeo Boxes in the Study of Development," *Science* 236 (1987): 1245–52.

33. See, for example, Weber, *Philosophy of Experimental Biology*, chap. 7; Philip Kitcher, "Genes," *British Journal for the Philosophy of Science* 33 (1982): 337–59; Richard M. Burian, "On Conceptual Change in Biology: The Case of the Gene," in *Evolution at a Crossroads: The New Biology and the New Philosophy of Science*, ed. David Depew and Bruce Weber (Cambridge, MIT Press, 1985), 21–42; Raphael Falk, "What Is a Gene?" *Studies in History and Philosophy of Science* 17 (1986): 133–73; Thomas Fogle, "Are Genes the Units of Inheritance?" *Biology and Philosophy* 5 (1990): 349–72; C. Kenneth Waters, "Genes Made Molecular," *Philosophy of Science* 61 (1994): 163–85; C. Kenneth Waters, "Molecules Made Biological," *Revue Internationale de Philosophie* 214 (2000): 539–64; Richard M. Burian, Robert C. Richardson, and Wim J. Van der Steen, "Against Generality: Meaning in Genetics and Philosophy," *Studies in History and Philosophy of Science* 27 (1996): 1–29; Peter Beurton, Raphael Falk, and Hans-Jörg Rheinberger, eds., *The Concept of the Gene in Development and Evolution: Historical and Epistemological Perspectives* (Cambridge: Cambridge University Press, 2000).

34. A classical complementation test works only with recessive alleles. However, there are a number of genetic tricks that can be used to identify genes with dominant alleles. See the Bithorax example discussed below.

35. Petter Portin, "The Concept of the Gene: Short History and Present Status," *Quarterly Review of Biology* 68 (1993): 173–223.

36. Weber, "Representing Genes," 296.

37. Portin, "The Concept of the Gene," 184.

38. Matthew P. Scott, "Complex Loci of *Drosophila*," *Annual Review of Biochemistry* 56 (1987): 195–227.

39. Kevin O'Hare et al., "DNA Sequence of the *White* Locus of *Drosophila melanogaster*," *Journal of Molecular Biology* 180 (1984): 437–55.

40. This term is due to Fogle, "Are Genes the Units of Inheritance?"

41. Scott, "Complex Loci of *Drosophila*," 223.

42. Jean-Noël Freund et al., "Molecular Organization of the *Rudimentary* Gene of *Drosophila melanogaster*," *Journal of Molecular Biology* 189 (1986): 4639–47; William A. Segraves et al., "The *Rudimentary* Locus of *Drosophila melanogaster*," *Journal of Molecular Biology* 175 (1984): 1–17.

43. Edward B. Lewis, "A Gene Complex Controlling Segmentation in *Drosophila*," *Nature* 276 (1978): 565–70.

44. Ernesto Sánchez-Herrero et al., "Genetic Organization of *Drosophila* Bithorax Complex," *Nature* 313 (1985): 108–13.

45. Peter A. Lawrence, *The Making of a Fly: The Genetics of Animal Design* (London: Blackwell, 1992), 214.

46. Atsushi Kuroiwa, Ernst Hafen, and Walter J. Gehring, "Cloning and Transcriptional Analysis of the Segmentation Gene *Fushi Tarazu* of *Drosophila*," *Cell* 37 (1984): 825–31; Amy J. Weiner, Matthew P. Scott, and Thomas C. Kaufman, "A Molecular Analysis of *Fushi Tarazu*, a Gene in *Drosophila melanogaster* That Encodes a Product Affecting Embryonic Segment Number and Cell Fate," *Cell* 37 (1984): 843–51.

47. The defining characteristic of such a series is that different alleles produce, in *trans* heterozygotes, a phenotype in which the phenotypic manifestations common to both alleles are expressed, whereas the traits with respect to which the two alleles show different manifestations are in the wild-type condition. Hence these alleles are nonallelic according to some traits (those traits which are normal in the heterozygote) and allelic for some other traits (those traits showing the mutant phenotype). These step-alleles can be arranged in a series ranging from alleles being almost fully nonallelomorphic with respect to each other (i.e., the heterozygous flies are almost normal) to alleles that seem to be allelomorphs in that the heterozygous show the full mutant phenotype. Because of this discrete array of degrees of allelomorphism, this phenomenon was termed "step-allelomorphism."

48. Elof A. Carlson, *The Gene: A Critical History* (Philadelphia: Saunders, 1966); Weber, "Representing Genes."

49. Antonio Garcia-Bellido, "Genetic Analysis of the *Achaete-Scute* System of *Drosophila melanogaster*," *Genetics* 91 (1979): 491–520.

50. Christine Dambly-Chaudière and Alain Ghysen, "Independent Sub-patterns of

Sense Organs Require Independent Genes of the *Achaete-Scute* Complex in *Drosophila* Larvae," *Genes and Development* 1 (1987): 297–306.

51. Sonsoles Campuzano et al., "Molecular Genetics of the *Achaete-Scute* Gene Complex of *D. melanogaster*," *Cell* 40 (1985): 327–38.

52. Alain Ghysen and Christine Dambly-Chaudière, "From DNA to Form: The *Achaete-Scute* Complex," *Genes and Development* 2 (1988): 495–501.

53. Maria C. Alonso and Carlos V. Cabrera, "The *Schaete-Scute* Gene Complex of *Drosophila melanogaster* Comprises Four Homologous Genes," EMBO *Journal* 7 (1988): 2585–91. This resolved an old puzzle surrounding the *achaete-scute* locus since the 1930s. While N. P. Dubinin and some other Russian geneticists had thought that the step-allelomorphism shown by mutants at this locus could be explained by the existence of several subgenes, Muller first argued that it contains three independent genes (*achaete, scute,* and *lethal of scute*). However, Muller later admitted that the locus was probably more complex than this. See N. P. Dubinin, "Stepallelomorphism in *Drosophila melanogaster*—the Allelomorphs Achaete2-Scute10, Achaete1-Scute11, and Achaete3-Scute13," *Journal of Genetics* 25 (1932): 163–81; H. J. Muller and A. A. Prokofyeva, "The Individual Gene in Relation to the Chromomere and the Chromosome," *Proceedings of the National Academy of Sciences of the United States of America* 21 (1935): 16–26; D. Raffel and H. J. Muller, "Position Effect and Gene Divisibility Considered in Connection with Three Strikingly Similar Scute Mutations," *Genetics* 25 (1940): 541–83.

54. Interestingly, Garcia-Bellido ("Genetic Analysis") inferred from his genetic data that the *achaete-scute* locus contains a "tandem repeat of reiterative signals" (491). This was spectacularly confirmed by the finding of two pairs of homologous transcription factors that seem to have arisen by gene duplication.

55. Waters, "Genes Made Molecular"; Waters, "Molecules Made Biological."

56. Kohler, *Lords of the Fly.*

57. Walter Gehring, in an interview of August 19, 2000, recalls that, inspired by the famous work of F. Jacob and J. Monod on gene regulation in bacteria, he contemplated the isolation of *Drosophila* analogs to DNA-binding proteins such as the *E. coli lac* repressor in the 1960s. However, this seemed impossible then because of the extremely low concentrations of such proteins in *Drosophila* embryos. Later on, attempts were made in Gehring's laboratory to isolate specific DNA-binding proteins. However, as Gehring anticipated, it would clearly be the approach taken for isolating genes such as *Antennapedia* by chromosomal walking that proved the most fruitful; see also Gehring, "Establishment of a *Drosophila* Gene Bank." In fact, this approach eventually led to the description of the protein-DNA interactions of homeodomain-containing *Drosophila* transcription factors at atomic resolution. See Martin Billeter et al., "Determination of the Nuclear Magnetic Resonance Solution Structure of an Antennapedia Homeodomain-DNA Complex," *Journal of Molecular Biology* 234 (1993): 1084–93.

58. Remarkably, the molecular researchers used a cytological map of the four *Drosophila* chromosomes prepared by Calvin B. Bridges in 1941.

59. Gehring, *Master Control Genes in Development and Evolution*, 49.

60. Hans-Jörg Rheinberger, *Toward a History of Epistemic Things: Synthesizing Proteins in the Test Tube* (Stanford: Stanford University Press, 1997), 135–36.

61. T. S. Painter, "A New Method for the Study of Chromosome Rearrangements and the Plotting of Chromosome Maps," *Science* 78 (1933): 585–86; T. S. Painter, "Salivary Chromosomes and the Attack on the Gene," *Journal of Heredity* 25 (1934): 465–76.

62. Joseph G. Gall and Mary Lou Pardue, "Formation and Detection of RNA-DNA Hybrid Molecules in Cytological Preparations," *Proceedings of the National Academy of Sciences of the United States of America* 63 (1969): 378–83; H. A. John, M. L. Birnstiel, and K. W. Jones, "RNA-DNA Hybrids at the Cytological Level," *Nature* 223 (1969): 582–87; Mary Lou Pardue and Joseph G. Gall, "Molecular Hybridization of Radioactive DNA to the DNA of Cytological Preparations," *Proceedings of the National Academy of Sciences of the United States of America* 64 (1969): 600–604.

63. Another striking example of such a hybrid technique is the approach of microdissecting polytene chromosomes mechanically and to amplify the DNA fragments dissected by inserting them into a cloning vector; see F. Scalenghe et al., "Microdissection and Cloning of DNA from a Specific Region of *Drosophila melanogaster* Polytene Chromosomes," *Chromosoma* 82 (1981): 205–16. In this case, the preparation technique used for cytological mapping becomes itself a source of specific DNA fragments (which can be as small as 200 kb) for cloning experiments.

64. Hans-Jörg Rheinberger, *Toward a History of Epistemic Things*, 135–36.

65. I am tempted, at this point, to hazard an analogy. Some theoretical concepts of classical physics have been invalidated by the quantum and relativistic revolutions, for example, the Newtonian concept of mass. Nevertheless, these concepts continue to be used for conducting measurements. In quantum mechanics, in fact, the concepts of classical physics are indispensable for measurements, even though these concepts are not applicable to quantum systems. By analogy, even though the classical concept of the gene was largely invalidated by advances in molecular biology, it continues to be used in experimental practice. I am not sure how seriously we should take this analogy. Clearly, in contrast to quantum mechanics (on the Copenhagen interpretation), the indispensability of the classical gene concept is not a matter of physical necessity.

Wormy Logic: Model Organisms as Case-Based Reasoning

It's a motley collection of creatures: They fly, swim, wiggle, scurry, or just blow in the wind. But to the scientific community, this compilation has been elevated above all other species. They are the model organisms.

—*Christine Bahls, Jonathan Weitzman, and Richard Gallagher, "Biology's Models"*

Although various strains of numerous laboratory organisms have proven biologically and historically significant, model organisms have become a cornerstone of research in the biomedical sciences, especially in the past few decades. In addition to the mapping and sequencing of the human genome, among key components of the Human Genome Project (HGP), which officially began in 1990, was the mapping and sequencing of the genomes of nonhuman model organisms, including mice, nematode worms, flies, *E. coli*, and yeast.[1] James Watson has described the idea to include nonhuman model organisms in the HGP as his most important contribution to the project.[2] Despite this sort of support from early enthusiasts, some of the more contentious issues raised during the preliminary planning stages of the HGP related to the model organism projects, concerning perhaps most importantly whether genetic sequencing was likely to result in knowledge relevant for the understanding and treatment of human disease processes, especially given the large amount of DNA without known function often derogatorily termed "junk DNA." Research on model organisms was rarely explicitly defended in the context of the project in its earliest days, perhaps in part because of assumptions about public and political perceptions and the lack of ability (or desire) to understand this research, despite the organisms' specific inclusion.[3] These organisms were used in the HGP as a means for developing the various mapping and sequencing technologies needed to study the more complex human genome, thus allowing the technologies to be tested and refined in a simpler, more efficient, and (purportedly) less expensive manner.[4]

But the genomes of these model organisms were also mapped and sequenced because they were expected to provide a basis for understanding normal gene regulation and human genetic disease, and more generally fundamental developmental, physiological, and other biological processes. Such expectations were based on the idea that many genetic and biological similarities exist between those organisms selected to serve as model organisms and humans; therefore model organisms would provide information that could aid in the interpretation of human genomic sequences and their products. This concept is rooted in the idea of the conservation of many mechanisms and processes: "Because all organisms are related through a common evolutionary tree, the study of one organism can provide valuable information about others. Much of the power of molecular genetics arises from the ability to isolate and understand genes from one species based on knowledge about related genes in another species. Comparisons between genomes that are distantly related provide insight into the universality of biologic mechanisms and identify experimental models for studying complex processes."[5]

Both the prevalence and centrality of model organisms in contemporary biomedical research, and claims about their use as the basis for deriving insight into certain common or even universal biological mechanisms, generate an ideal laboratory for the examination of epistemic issues related to the use of such organisms. In addition, the growing literature within the history and philosophy of science on conceptual issues associated with modeling and representation in science,[6] and on various model organisms,[7] creates a space within which close attention to the principles and practices associated with such models may prove fruitful.

This essay examines the conceptualization of model organisms as models, and presents a formal account of how they are used to generate knowledge through what can be viewed as a form of case-based reasoning. Following a brief historical account of the development and use of one model organism, the nematode worm *Caenorhabditis elegans*, I address questions about the methodologies underlying the work on genetic sequencing and developmental processes in this organism. In particular, I ask: What types of reasoning ground the use of experimental organisms when they are being developed and used as model organisms, and how are these models refined over time?

Some clarifications on terminology to begin: the term *model organism* is used throughout this essay rather than *model system* since the former expression is explicitly employed in the literature on the HGP, and more generally in contemporary organism-based biology.[8] Model organisms can be seen as a specialized subset of the more general class of model systems, where the latter usually encompasses not only the organism but also the techniques and experimental methodologies surrounding the organism itself.[9] This essay explores some of the techniques and methods used to establish and refine model organisms, yet primarily from the point of view, as it were, of the model organisms themselves.

BACKGROUND: THE WORM

C. elegans is a free-living nematode, around a millimeter in length, with extremely simple behaviors and structures and a relatively recent history as a model organism.[10] As noted in the Nobel Prize for Physiology or Medicine presentation speech for 2002, which celebrated three worm workers and the "joy of worms," part of what makes it a good candidate for a model organism is that *C. elegans* is "loaded with features."[11] There are two sexual forms, a self-fertilizing hermaphrodite and a rarer male that can fertilize hermaphrodites, which differ slightly in appearance and structure; this feature makes it an excellent genetic system as organisms can either be purebred by isolating hermaphrodites, or new genetic material can be introduced via breeding with males. The adult is composed of a tube made of an exterior cuticle, which contains two smaller tubes (the pharynx and the gut) and the reproductive system. The organism is transparent throughout its life cycle, making observation of many biological processes possible by various forms of microscopy. The genome of *C. elegans* is approximately 100 million base pairs, one-thirtieth the size of the human and twenty times that of *E. coli*, and it was virtually completely sequenced as of December 1998.[12]

The choice of *C. elegans* by Sydney Brenner in the mid-1960s and the original pursuit of research focused on this organism primarily at a single institution (the Laboratory of Molecular Biology in Cambridge, England) to which most current-day researchers can trace their own lineages has resulted in a relatively cohesive community often celebrated as a model of scientific cooperation and shared understanding of fundamental concepts.[13] Hence an analysis of how "the worm" (as it is called by researchers in this area and many in the broader scientific community) functions as a model organism can be used as the basis for understanding the epistemic structure underlying most ongoing research in this area.

A general examination of the history of organism choice reveals that prospective model organisms are typically selected and constructed based not mainly on principles of or knowledge about the universality or even typicality of their biological characteristics and processes (though it is hoped that many features will prove to be shared by or common to other organisms), but primarily due to perceived experimental manipulability and tractability. For example, *C. ele-*

gans was chosen specifically for its developmental invariance and simplicity, despite the atypicality of these biological characteristics (among many others) of *C. elegans*, even in comparison to other closely related organisms. In short, the general aim of the original research project was to achieve an understanding of developmental processes in metazoans (animals with bodies composed of differentiated cells, as opposed to protozoa or unicellular animals), and in particular, the development of the nervous system, since it was thought to be the most complex and interconnected system in these organisms.[14]

Brenner wanted to do research with an organism that proved experimentally straightforward to manipulate and had relatively basic behaviors and structures, but was not so simple as to be "unrepresentative." The goal was to optimize an organism, in large part through making a careful organismal choice to start, rather than focusing on achieving standardization once in the laboratory via inbreeding and other typical techniques. Brenner and most subsequent worm workers in the early years of the research implicitly assumed that although *C. elegans* is simple, it is similar to all (or most) of the more complex members of the metazoa in terms of the genetic control of cellular differentiation. In particular, the genetic control of the development of the structure of the nervous system was thought to be likely to have shared fundamental mechanisms, in large part because of an implicit assumption of genetic conservation, particularly of essential processes.[15]

One of the primary ways in which *C. elegans* can be seen as a model organism relies on the idea that a model has been established to which particular empirical instantiations (i.e., actual material worms) can be compared in order to articulate variations and differences in various features. The use of this form of reasoning is perhaps most familiar from basic genetics: the first step in the underlying strategy is to select and establish a so-called wild type for the organism (taken as a standard from among other possible wild types available in nature) against which other genetic variants or abnormal types can be compared. Despite its name, the wild type may not be the most common, frequent, or even a "normal" version of the organism: sometimes it is simply the first strain discovered on which subsequent research has been based, but oftentimes it is the easiest to manipulate experimentally. These experimental organisms of course are natural inasmuch as they are still actual, living, concrete organisms that have been "selected from nature's very own workshop."[16] However, the carefully selected wild type is, in this sense, an idealized model of actual organisms in nature since the latter oftentimes end up differing considerably from those highly rarified beasts that remain isolated in the laboratory, particularly as a model organism comes to be more widely used.[17] Thus modeling occurs most obviously in the establishment of the wild type, an essential first step to establishing and using something on an ongoing basis as a model organism. Without this process, it is impossible to have a norm against which "abnormal" (or more precisely, that which is variant) can be compared in terms of genetics, developmental lineages, and so on. So a worm abnormal in movement might be detected by comparison of the paths it traces in response to a stimulus to those traced by a worm held to be normal.

A second way in which modeling occurs is in the establishment and use of what I have called elsewhere a "descriptive model."[18] The term *descriptive* is utilized to capture the idea that these sorts of models are descriptions that serve as prerequisites to explanatory questions; their articulation often is not motivated (at least immediately) by their future potential explanatory value. Thus in model organism work, an extensive research phase is usually dedicated to developing a descriptive model of the organism. Consider, for example, the articulation of the so-called wiring diagram of the neural connections within *C. elegans*. This model was a paper (and later computerized) series of drawings that resembled electric circuitry diagrams.[19] The overall diagram was constructed by combining wiring diagrams from several individual wild-type worms, not only because of practical or experimental limitations but because it was deemed necessary to eliminate what seemed to be individual neural differences (even between genetically identical organisms) in favor of a canonical nervous system. The wiring diagram is based on an abstract model of the worm in terms of the typical or usual neural connections exhibited not by any one specimen alone, or by numerous individual organisms, but by a more abstract construct hybridized from a few individual specimens. The wiring diagram thus is a model of the worm in terms of the typical or usual neural connections exhibited not by any one specimen taken by itself but by a very precisely derived type of construct.[20] This descriptive model is compared to the wiring diagrams for worms that are variant or abnormal in neural patterns in order to assess possible connections between variations in genetic sequence and in neural structure, and eventually to test the range of the applicability of the descriptive model. Thus in this sense, some aspects of model organisms are in fact more like mechanical or physical models, constructed from natural organisms but constructed nonetheless, and hence highly idealized since individual differences among wild-type worms have been eliminated in lieu of (what are thought to be) the most commonly occurring structures.

Laboratory and community practices thus allowed the articulation and refinement of *C. elegans* as a model organism through at least two forms of idealization: the choice of a wild type (which provides concrete laboratory instantiations of the organism, permitting comparison, for instance, to particular mutant strains) and data-summarizing descriptive devices (such as wiring or cell lineage diagrams). Diverse model organisms undoubtedly have different histories that involve various kinds of modeling, depending on the natural features of the organism targeted to be exploited, the goals of the research community, and the degree of development of the model organism, among other factors.

THE PRINCIPLES OF CASE-BASED REASONING

The biomedical and human sciences have a long history of the use of the case study as an object through which knowledge is generated and phenomena are made intelligible. The case is used to capture or summarize clinical and empirical data, to investigate underlying theories of disease, and communicate findings to other practitioners and researchers, among other purposes. To begin, it is helpful to provide a brief overview of the general form of case-based reasoning as used in medicine and elsewhere.[21] The basic method proceeds by construction of what might be termed (borrowing from the language of artificial intelligence) an "index case," in more or less detail depending on the goals of the situation in which it is to be used. In medicine, for instance, the index case often begins as a syndrome letter or published report on an individual patient, which then is abstracted into a model for more general use.

The case's use occurs through retrieval when a practitioner is presented with a new case that seems to have some overlap with the original index case, at least in terms of the details believed most relevant. This process involves a form of separating signal from noise, to put it in different terms. The result is a feedback loop between processes of justification of the fit between the original index case and the new case under examination, particularly via the assessment of similarity and identity relations. The outcome might not only be pragmatic (i.e., it may provide the basis for making a diagnosis or prognosis) but in addition, new cases can lead to the modification of the index case as appropriate over time, or even to the adoption of a new index case for a particular condition, which in turn is disseminated through publication and teaching.

Underlying the index case and the feedback loop between it and any new instances is an even more basic index case: that of the human being who is "normal" with respect to the abnormal features noted in the index case. What is considered to be the index case for the normal (i.e., the undiseased condition) may also be altered over time as the range of variants or errors in what were assumed to be the shared or common attributes (genetic, physiological, and otherwise) among healthy individuals are discovered. Thus the index case of the normal and of the disease condition often are constructed (and reconstructed) in terms of each other as more knowledge is gathered. What is essential in this form of reasoning is the feedback loop that exists between the descriptive model of the normal and the descriptive model of the abnormal condition. Newly acquired evidence can change what is considered to be the index case or whether something should be considered a unique case at all.

Thus these cases are models inasmuch as that although they originate from some actual observed instance in the first place, once they begin to be disseminated and used, they become idealized away from particular details of the observed phenomena. They serve as intermediaries between the base of available knowledge (which is oftentimes overwhelmingly descriptive and relatively lacking in formal theoretical structure) and new natural phenomena that present themselves and require understanding or explanation. Thus the following types of cases can be viewed as models, in the sense advocated by Mary Morgan and Margaret Morrison (among others): they cannot be derived from either theory or data, and hence are partially independent; they clearly mediate between theory and the world and are used in a tool-like manner to perform a range of tasks.

What is most important to note here is that as the index case is refined over time, a tension arises: in some sense, the base index case comes closer to what is really out there in nature, while at the same time it becomes more distant from any one concrete individual instantiation (any actual, material organism). Nonetheless, it remains a model, fulfilling many of the attributes that we expect from models: it is idealized, in that no patient typically fulfills all of the conditions captured in the model, and yet patients can still be identified as having a condition or being an instance of that particular disease category or case.

MODEL ORGANISMS AS CASES

The practices of contemporary biological science have (potentially conflicting) goals similar to those found in the practice of the medical sciences. There is a desire to get to the fundamental biological characteristics shared by all living things, be they biochemical, genetic, developmental, or neurobiological processes. At the same time, biologists are aware that any model system or organism selected for research may be problematic and atypical, particularly inasmuch as such systems are proving to be complex in ways previously not anticipated. The previous section on *C. elegans* as a model organism has shown several ways in which the organism as studied by biologists constitutes an idealized entity or a model. The epistemic strategy of using the models as cases allows them to serve as a means of control of complexity, as a way to create an appropriately simplistic yet descriptively rich basis for future studies and for more traditional hypothesis testing, experimentation, and explanation.

Different aspects of a model organism can thus be viewed as index cases on which comparison to variant and abnormal instances of the same organism can take place. So, for instance, the wild type of the natural organism serves as an index case in that it establishes a genotype that comes to be understood as normal and serves as the basis for comparison to subsequent cases of abnormal or variant genotypes. Similarly, the wiring diagram captures another sort of basic index case to which variations in neural structure can be compared. Among the key foundational assumptions used to determine what counts as the relevant or most useful base index case for an organism are the anticipated degree of genetic homology and genetic conservation. Eventually the index case may be altered to better reflect an increased understanding of what is generalizable (or perhaps universal, at least within the species) in the model organism.

There are at least two important points implicit in this account supporting the claim that the types of models discussed serve as the basis for case-based reasoning processes using model organisms. First, the idea that model organisms are in fact idealized entities as outlined has resonance with the construction of epistemologic entities elsewhere in the sciences, for instance, of the "average man" in medical and human sciences, going back to the work of Adolphe Quetelet: "The consideration of the average man is so important in the medical sciences that it is almost impossible to judge the state of an individual without comparing him to a fictive being that one regards as being the normal state and who is nothing but the [average man]."[22] Rich, descriptive idealizations thus constitute the starting point for case-based reasoning as some baseline case must be provided to initiate the reasoning process. Yet these idealized cases necessarily remain fictitious, as does the nervous system of the so-called canonical worm, at the same time as they constitute essential tools for developing an understanding of the actual organism.

Second, note that as with medical case reports, usually there is no explicit (or implicit) testing of a hypothesis or theory, or what might be considered other typical scientific behaviors. Instead, the process proceeds by the proffering of observations and detailed descriptions, which may well point to testable hypotheses and explanations, particularly if they are to have an impact on the development of theory or on practice.[23] Thus there is a creation of an epistemological space or framework within which to ask questions. However, as bluntly stated by a commentator on medical reasoning, "with higher organisms, and especially with patients, it becomes hopeless to attempt to create complete descriptions. . . . This is a kind of epistemologic surrender and consists in simply ignoring many of the things that could be truthfully said in order to say what must be said."[24] Both in medicine and in biological reasoning from model organisms, complexity, completeness, and perhaps "naturalness" are sacrificed in favor of the selective construction of manageable material and a framework within which scientists can work and ask questions.

Case-based reasoning using descriptive models in biology thus relies on a double feedback loop between an index case and a case of interest that is abnormal or variant in some way. As in medicine, various cases are developed, for instance, composed of descriptions of genetic or neural sequences in organisms. The base index case begins with a descriptive model of the organism established as being normal in phenotype, for which, say, the genomic sequence is identified and established as normal (or at least a norm against which other cases can be measured). This sequence can then be compared to that of organisms abnormal in phenotype (and thus assumed to be abnormal in genetic sequence) in order to draw out the functional properties of the genomic sequence within the particular model organism. Furthermore, an additional level of case-based reasoning occurs which then holds that determining the sequences in a variety of model organisms will reveal conserved (normal) genetic regions, which in turn will allow an investigation of the same part of the sequence in the normal human genome (or other "higher" organisms) and prove fruitful for understanding the functional properties of these sequences. Finally, the eventual goal is to understand the higher-level, phenotypic results of abnormal, human genomic sequences found to be similar to the base case, namely, the "abnormal" (or variant) sequences in the model organism, based on a correlation between these sequences and higher-level properties such as disease conditions or other abnormalities.

CONCLUSIONS

What is most important to notice when analyzing the use of model organisms, and particularly the way in which they function as a form of case-based reasoning, is that answering the question of whether a model organism will in fact prove a useful model (i.e., for human genome sequencing) requires that researchers not only work on sequencing in the model organism but that this sequencing occur in tandem with sequencing in the object of interest, the human genome, and other comparative genomic work. This conclusion points to an important, but easily overlooked, aspect of modeling: in order for models to actually function well as models, an ongoing refinement of the original descriptive models (the base index cases) must occur. So must a constant interplay between the original descriptive model and the subject modeled (or the cases of interest or targets), and the continuous development of the positive analogies between them (along with an identification of the relevant disanalogies and their import).[25] Much rhetoric surrounding model-organism research unconstructively obscures this interplay and hence misrepresents the potential limitations of even good models. In other words, providing a model requires an interaction between the model and the object of interest being modeled, or between the base index case and the case of interest, including the construction of similarity relations, which are impossible to devise without a detailed description of the process to be modeled (which in this case includes the functional properties of the sequence).

Case-based reasoning is an epistemic process that is far from straightforward, and it may seem to fail to allow us to obtain the usual results we expect in science inasmuch as it fails (at least initially) to produce unified theories or mechanistic explanations, instead resulting in a form of scientific understanding (perhaps of a weaker sort than our traditional theories and explanations) that is constantly evolving, incomplete, and uncertain, but nonetheless has the status of knowledge for its practitioners. Model organisms and their features that serve as cases mediate between theory and the world (and cannot be derived directly from either data or theories) and come to be used in a tool-like manner to perform a range of tasks, perhaps the most important of which is establishing a framework within which to ask questions.

NOTES

I am grateful to the organizers and participants at the original Princeton workshop at which an earlier version of this essay was presented for constructive feedback, particularly to Angela Creager and Mary Morgan, as well as numerous other colleagues who have patiently supported and provided me with feedback on the various incarnations of this research project. My research was partially conducted while I was a visiting fellow/faculty member at the Shelby Cullom Davis Center for Historical Studies at Princeton University, as well as while I was teaching in the Department of Economic History, London School of Economics, as part of the Leverhulme, UK, Economic and Social Research Council sponsored project "How Well Do Facts Travel?" I also wish to thank my research assistant Fiona Mackenzie for sourcing literature.

1. See Rachel A. Ankeny, "Model Organisms as Models: Understanding the 'Lingua

Franca' of the Human Genome Project," *Philosophy of Science* 68 (2001): S251–S261.

2. James D. Watson, "The Human Genome Project: Past, Present, and Future," *Science* 248 (1990): 44–48; James D. Watson, "A Personal View of the Project," in *The Code of Codes: Scientific and Social Issues in the Human Genome Project*, ed. Daniel J. Kevles and Leroy Hood (Cambridge: Harvard University Press, 1992), 164–73. See also Roger Lewin, "The Worm at the Heart of the Genome Project," *New Scientist* 127 (1990): 38–42.

3. See, for instance, the discussion in Charles R. Cantor, "Orchestrating the Human Genome Project," *Science* 248 (1990): 49–51.

4. U. S. Department of Health and Human Services and U. S. Department of Energy, *Understanding Our Genetic Inheritance: The U. S. Human Genome Project; The First Five Years, Fiscal Years 1991–1995* (Washington: Government Printing Office, 1990). Note, however, that many model-organism researchers participated in the HGP in large part to be able to study their organisms of choice in their own right, which in turn created various epistemic and pragmatic tensions within many laboratories and research programs, a point I cannot examine in any detail here.

5. Francis S. Collins et al., "New Goals for the U.S. Human Genome Project: 1998–2003," *Science* 282 (1998): 686–87.

6. For instance, see Mary S. Morgan and Margaret Morrison, eds., *Models as Mediators: Perspectives on Natural and Social Science* (Cambridge: Cambridge University Press, 1999); Soraya de Chadarevian and Nick Hopwood, eds., *Models: The Third Dimension of Science* (Stanford: Stanford University Press, 2004).

7. Though by no means an exhaustive list, historical and conceptual accounts of the development of and research with various model organisms that have influenced my research include Adele E. Clarke and Joan Fujimura, "What Tools? Which Jobs? Why Right?" in *The Right Tools for the Job: At Work in Twentieth-Century Life Sciences*, ed. Clarke and Fujimura (Princeton: Princeton University Press, 1992), 3–44, as well as the other essays in this collection; Richard M. Burian, "How the Choice of Experimental Organism Matters: Epistemological Reflections on an Aspect of Biological Practice," *Journal of the History of Biology* 26 (1993): 351–67, as well as the other articles contained in this special issue of the journal devoted to experimental organisms; Robert E. Kohler, *Lords of the Fly: Drosophila Genetics and the Experimental Life* (Chicago: University of Chicago Press, 1994); Angela N. H. Creager, *The Life of a Virus: Tobacco Mosaic Virus as an Experimental Model, 1930–1965* (Chicago: University of Chicago Press, 2002); Karen A. Rader, *Making Mice: Standardizing Animals for American Biomedical Research, 1900–1955* (Princeton: Princeton University Press, 2004); and PhD research in progress by Sabina Leonelli (Vrije Universiteit, Amsterdam) on *Arabidopsis thaliana*.

8. Compare Jessica A. Bolker, "Model Systems in Developmental Biology," *BioEssays* 17 (1995): 451–55.

9. On model systems, see, for instance, Hans-Jörg Rheinberger, *Toward a History of Epistemic Things: Synthesizing Proteins in the Test-Tube* (Stanford: Stanford University Press, 1997).

10. For additional background on the history of the choice and use of *C. elegans*, see Rachel A. Ankeny, "The Conqueror Worm: An Historical and Philosophical Examination of the Use of the Nematode *C. elegans* as a Model Organism" (PhD diss., University of Pittsburgh, 1997); Soraya de Chadarevian, "Of Worms and Programmes: *Caenorhabditis elegans* and the Study of Development," *Studies in the History and Philosophy of Biological and Biomedical Sciences* 29 (1998): 81–105; Rachel A. Ankeny, "The Natural History of *C. elegans* Research," *Nature Reviews Genetics* 2 (2001): 474–78; Kenneth F. Schaffner, "Genetic Explanation of Behavior: Of Worms, Flies, and Men," in *Genetics and Criminal Behavior,* ed. David Wasserman and Robert Wachbroit (Cambridge: Cambridge University Press, 2001), 79–116.

11. Urban Lendahl for the Nobel Committee, Karolinska Institutet, Stockholm, December 10, 2002, available at nobelprize.org/nobel_prizes/medicine/laureates/2002/presentation-speech.html.

12. *C. elegans* Sequencing Consortium, "Genome Sequence of the Nematode *C. elegans*: A Platform for Investigating Biology," *Science* 282 (1998): 2012–18; for general overviews of work on this organism, see William B. Wood and the Community of *C. elegans* Researchers, eds., *The Nematode Caenorhabditis elegans* (Cold Spring Harbor, N.Y.: Cold Spring Harbor Laboratory Press, 1988); Donald L. Riddle et al., eds., *C. elegans II* (Cold Spring Harbor, NY: Cold Spring Harbor Laboratory Press, 1997).

13. Leslie Roberts, "The Worm Project," *Science* 248 (1990): 1310–13; E. Pennisi, "Worming Secrets from the *C. elegans* Genome," *Science* 282 (1998): 1972–74.

14. Sydney Brenner, foreword to Wood and Community of *C. elegans* Researchers, *The Nematode,* ix–x.

15. The idea of shared mechanisms can be taken to its extreme: Howard Gest has suggested that the literature surrounding the current proliferation of model systems (or organisms, to use my preferred terminology) often seems to use *model* to signify universality, and has called for a correction of what he considers linguistic misusage (which I would claim actually has much deeper, epistemic implications); see Howard Gest, "Arabidopsis to Zebrafish: A Commentary on 'Rosetta Stone' Model Systems in the Biological Sciences," *Perspectives in Biology and Medicine* 37 (1995): 77–85.

16. Evelyn Fox Keller, *Making Sense of Life: Explaining Biological Development with Models, Metaphors, and Machines* (Cambridge: Harvard University Press, 2002), 51.

17. In organisms in which there is ongoing flow over time between the laboratory and the field or the wild, the amount of idealization in the model may be reduced, or more precisely, there may be more than one strain or variant held as a norm.

However, particularly with genetic model organisms (those selected primarily because of their power for genetic analysis, which is my focus in this essay), it is essential to settle on (and persist in using) one wild type.

18. See Rachel A. Ankeny, "Fashioning Descriptive Models in Biology: Of Worms and Wiring Diagrams," *Philosophy of Science* 67 (2000): S260–S272. I am extremely grateful to Sabina Leonelli for her helpful critique of my overemphasis in this earlier article on the abstract features of *C. elegans* as model organism, due in part to my examination concerning itself solely with the construction of the worm's wiring diagram; I have attempted to clarify and remedy this narrow focus in the current essay.

19. John G. White et al., "The Structure of the Nervous System of the Nematode *Caenorhabditis elegans*: The Mind of a Worm," *Philosophical Transactions of the Royal Society of London: Series B, Biological Sciences* 314 (1986): 1–340.

20. This account has resonance with Jim Griesemer's analysis of material model building inasmuch as the wiring diagram (as well as cell lineage and other descriptive models associated with model organisms) can be seen as serving as "vicarious" models that serve as the basis for future theory development; see Jim Griesemer, "Material Models in Biology," in *PSA 1990*, ed. Arthur Fine, Micky Forbes, and Linda Wessels (East Lansing, MI: Philosophy of Science Association, 1991), 2:79–93.

21. This discussion summarizes a more detailed examination by me in "Case-Based Reasoning in the Biomedical and Human Sciences: Lessons from Model Organisms," in *Logic, Methodology and Philosophy of Science: Proceedings of the Twelfth International Congress*, ed. Petr Hájek, Luis Valdés-Villanueva, and Dag Westerståhl (London: King's College Publications, 2005), 229–42. On case-based reasoning in the human sciences, see especially John Forrester, "If *p*, Then What? Thinking in Cases," *History of the Human Sciences* 9 (1996): 1–25.

22. Adelphe Quetelet, *Sur l'homme et le développement de ses facultés, ou essai de physique sociale* (Paris: Bachelier, 1835), 2, 267, as quoted in Jonathan Cole, "The Chaos of Particular Facts: Statistics, Medicine and the Social Body in Early Nineteenth-century France," *History of the Human Sciences* 7 (1994): 12.

23. See R. J. Simpson and T. R. Griggs, "Case Reports and Medical Progress," *Perspectives in Biology and Medicine* 28 (1985): 402–6.

24. M. S. Blois, "Medicine and the Nature of Vertical Reasoning," *New England Journal of Medicine* 318 (1988): 848.

25. Note the resemblance of this to views on models from the classic book by Mary B. Hesse, *Models and Analogies in Science* (London: Sheed and Ward, 1963).

Model Organisms as Powerful Tools
for Biomedical Research

E. JANE ALBERT HUBBARD

In the past several decades, the marriage of genetics and molecular biology has produced an approach to basic biological research that exemplifies science in its purest form: a response to the call of curiosity aroused by the perception of reality. Model organisms constitute the tools and subjects of this approach.[1] These studies are often begun without certainty as to the future use or merit of the work, yet their results are poised to revolutionize medicine. Initially, the public agencies in the United States that fund biomedical research appeared reluctant to support such supposedly risky undertakings. Now, in the face of the potential gains from model organism studies, the understanding has emerged that it constitutes perhaps a greater risk not to fund this type of research. Permitting fertile and inquisitive minds to follow their fascination offers a path full of potential benefit beyond the apparent short-term goals of the research. Indeed, model organisms were used in work recognized by the 1995 Nobel Prize in Physiology or Medicine to Edward B. Lewis, Christiane Nüsslein-Volhard, and Eric F. Wieschaus "for their discoveries concerning the genetic control of early embryonic development," and again by the 2002 prize to Sydney Brenner, H. Robert Horvitz, and John E. Sulston "for their discoveries concerning 'genetic regulation of organ development and programmed cell death.'"[2] Moreover, the fact that many major pharmaceutical companies now employ model-organism research strategies stands as a testament to the applicability of this research to medicine.[3]

In this article, I hope to provide a glimpse of the astonishing utility of model organisms as tools in biomedical research.[4] First, I will introduce the concept of a model organism (what it is and what it is not). Second, I will introduce my favorite model organism, the worm *Caenorhabditis elegans*. Finally, I will relate several stories to illustrate more clearly the impact of this type of research on medicine.

MODEL ORGANISMS IN GENERAL

What is a model organism? Is it a model of an organism? Is it the archetype? No. The term *model organism* refers to an organism used to gain an in-depth understanding of a biological problem. This definition may appear vague, but this results from the versatile nature of the model-organism approach to biological problems. The basic strategy of model-organism research is this: starting with a phenomenon of interest to the investigator, the investigator chooses an organism that permits a great depth of understanding of the phenomenon. Other strategies have followed from this starting point. Recently, workers have converged on a handful of relatively simple creatures in which a combination of genetics, molecular biology, biochemistry, and genomics can be applied to unlock astonishing mysteries of life. This strategy alone, however—an interesting phenomenon and an organism in which to study it in detail—could lead to nothing more than a specific comprehension of many isolated and insignificant phenomena in obscure organisms. Interesting, yes, but not the stuff of biomedical revolutions.

What makes these organisms so powerful as biomedical research models is the fact that genetic and molecular functions are conserved across species. The conservation of genetics is what allowed Gregor Mendel's famous studies with the garden pea to be regarded as laws. Both pea and human chromosomes assort independently during meiosis. Furthermore, the actual genetic code that translates DNA sequence into the amino acid sequence proteins is, astonishingly, conserved among all living things.[5]

The conservation of molecular function means that the proteins encoded by the DNA of one organism will usually act similarly in chemical reactions in the cells of all organisms. Consequently, human DNA (encoding a specific gene) that is spliced into the chromosome of a worm will usually encode the correct protein and the protein will perform essentially the same chemical task in a worm cell as it would have performed in the human cell (albeit often with different consequences—this will be clarified below). Finally, the accumulation of more and more DNA sequence data from diverse groups of organisms identifies families of functionally similar proteins, and their sequences can be compared. These data, in turn, help experimentalists assess protein structure-function relationships within and between species.[6]

Still, what do model organisms and genetic and molecular conservation have to do with biomedical research, the object of which is to ameliorate human suffering by providing treatments and information for prevention? A modern corollary to this objective is that a detailed molecular understanding of the basis of disease will lead to better prevention and more efficacious treatments. Until fairly recently, animals in biomedical research were used strictly as disease models. A good disease model is an animal in which a human disease can be faithfully simulated. Good animal models exist for many diseases and have been used to understand the etiology, progression, and treatments of these diseases. Genetic experiments, especially in mice, take this concept a step further: rather than giving a mouse a particular infection, for example, a mutant mouse with a genetic predisposition to a certain disease or condition can be propagated for further study. With the newly published mouse genome, additional tools will be added to this arsenal.[7] Mice provide laudable approaches, but these approaches have limitations. For example, many human diseases have no animal-model equivalent, and despite their relatively closely shared ancestry with humans, even mouse disease models are limited by the physiological differences between mice and humans. In addition, despite tremendous advances in the genetic and molecular manipulation of mice, practical limitations such as the high cost of animal maintenance, relatively small brood sizes, and long generation time preclude certain experiments and screens possible using simpler model organisms.

In contrast to disease models, the model-organism approach to biomedical research is far broader, more flexible—and unpredictable. Clearly, model systems such as yeast, fruit flies, and worms are evolutionarily and physiologically even further removed from humans than mice, and therefore they would appear to serve as poor biomedical models. Because diseases are now being explored at the genetic, cellular, and biochemical levels, however, the conservation of genetic and molecular function decreases the relevance of this limitation. Certainly, there are diseases and conditions resulting from complex interactions particular to human physiology, and they will not prove amenable to the model-organism approach, but if the current data trend continues, these will be fewer than previously imaginable on the basis of morphological differences alone. Furthermore, even in cases where the effects of biochemical pathways are radically different, or the pathways are dissimilar, the model organism provides initial access to the inquiry. Often studies can be more easily extended in mammalian systems once a genetic, molecular, or biochemical foothold has been gained using simpler model organisms. Therefore the model organism can be used to probe deeply into the function of genes, proteins, and complexes even though the model organism itself does not manifest any symptoms similar to human disease. Hopefully, the illustrations below will clarify these claims.

To understand the myriad ways that model organisms are used in biomedical research and to appreciate the stories I will recount, the logic of the genetic approach to biological problems must be introduced. The "forward" genetic approach begins with a phenomenon and asks: What mechanisms are necessary for this phenomenon to occur normally? Take, for example, the question of what mechanisms are responsible for the formation of a fruit fly wing. The existence of genes necessary for the formation of normal wings was deduced from the discovery of mutant flies forming abnormal wings—or no wings at all. Many fruit fly mutants were discovered in the early decades of the twentieth century and were used to elucidate fundamental mechanisms of heredity.[8] More recently, molecular biology techniques permit these genetic results to be taken further by revealing the chemical basis for the mutant form of a gene (by cloning the gene originally identified by the very detection of the mutant fly and sequencing the corresponding mutant DNA). With these genetic and molecular results, several important inferences can be made. First, the fact that a specific change in the DNA sequence produces a specific morphological defect (say, an abnormal wing) means that the product (usually a protein) normally encoded by the gene plays a role in the normal process of wing formation. Second, since a large number of gene sequences are now in the databases, many proteins fall into recognizable families and these relationships often disclose the chemical function of the protein. Hence the correspondence between gene, protein, and morphology can be scrutinized *via* biochemical analysis. Finally, the actual mutation in the DNA that leads to a corresponding change in the protein (and, ultimately in morphology) can be found. This information often sheds more light on the biological process by revealing particular structure-function aspects of the protein involved.

The "reverse" genetic strategy begins with a gene or protein already known to have an interesting function (for example, a gene that when mutated is known to cause disease in humans). The function of the gene or protein is then manipulated in the model organism and the consequences are evaluated. Powerful genetic and molecular extensions of these so-called forward and reverse approaches, together with genome-level analyses, allow the identification of functionally related genes, proteins, and biochemical pathways in model organisms and in humans.[9]

A PARTICULAR MODEL ORGANISM

Model-organism research is moving forward in several species. Each model organism offers experimental advantages and disadvantages since the biology of the species and the history of its use as a model are varied. In fact, the evolutionary divergence of the model organisms has, itself, led to important discoveries when comparisons are made between them. Here, I will focus on work using the worm *Caenorhabditis elegans* as a model organism. This by no means tries to suggest that other model organisms are less valuable. The arsenal of tools and strengths differs among the organisms, making them all valuable; the same general strategies I explain here are employed with other organisms with equally amazing results.

The rise of *C. elegans* as a model organism can be traced to seminal work by Sydney Brenner, the results of which were published in 1974.[10] Brenner applied the forward genetic approach to the study of behavior and morphology in this tiny nematode worm (*C. elegans* is a microscopic worm about one millimeter in length that is normally found in soil). Brenner was interested in understanding the connection between genes and behavior, and he deemed the genetically well-described fruit fly as too complex for this task; Brenner wanted a simpler organism in which the complete structure of the nervous system could be determined. Brenner carefully chose this species for attributes that would give him access to the details he sought.[11] These animals are easily grown and manipulated in the laboratory, they can produce large numbers of progeny in a short time (a single worm can give rise to about one hundred thousand grand-progeny in one week—a key feature for the forward genetic approach), and they are virtually transparent under high magnification. This last feature allowed investigators to trace the lineage of each of the worm's ~1000 cells from the one-cell stage to two, four, eight, and so on to the adult.[12] In a culmination of many years of effort aided by great advances in molecular biology, the entire DNA sequence of this worm was published in 1998—the first multicellular organism for which the whole genome was sequenced.[13] It is safe to say this microscopic worm is known in greater anatomical, genetic, and molecular detail than any other animal on earth.

As I mentioned earlier, using a genetic approach, mutations can be sought that affect any phenomenon, so that provided the phenomenon is under genetic control, the genes important for the normal manifestation of that phenomenon can eventually be found. Since Brenner was looking for the connection between behavior and genes, he looked for mutants that could not behave (move) properly. In practice, this was done by soaking a population of worms in a chemical known to induce mutations in the DNA (this can also be accomplished by treating them with radiation like x-rays or ultraviolet light) and then looking among the progeny of the treated worms for animals with abnormal behavior. Since the entire cell lineage is known, the effect of the mutations on neurons or muscles, for example, can be studied separately. Finally, molecular biology techniques provide the means to manipulate DNA. Many of the mutations that

Brenner found are now understood at the level of the changes in DNA sequence and the altered protein products that result from these changes.[14]

In the thirty years since Brenner's publication, this lowly worm has contributed enormously to many different areas of biomedical science including oncology, neurological diseases, infectious disease, and pharmacology. This is possible because of the conservation of genetic and molecular function across species that I mentioned above. Although Brenner was originally interested in a model to study the genetic basis of behavior, it soon became clear that the utility of a model organism is limited only by the curiosity and imagination of the investigator. There are now over 1,200 researchers in approximately 450 laboratories using C. elegans as a model organism to study myriad biological processes.[15]

To illustrate some of the points I have made thus far, I will recount my own introduction to C. elegans research and describe results relevant to one C. elegans gene. Finally, I will provide several additional examples that demonstrate the versatility of this model organism and the ingenious ways it is being used to unravel mysteries of interest to human welfare.

MY PERSONAL ENCOUNTER WITH C. ELEGANS

Although I have been fascinated by the diversity of animal life since I can remember (and was particularly fascinated by developmental biology in high school), I first encountered the model-system concept in a developmental biology class as an undergraduate in 1980. By this time, scientists recognized the possibility that cancer resulted from mutations in genes normally present in the cell where they govern important cell functions and fate decisions. The discovery that cancer was linked to changes in normal cellular genes earned Michael Bishop and Harold Varmus the Nobel Prize in Physiology or Medicine nine years later. By 1980, the molecular revolution was underway, and the conservation of genetic and molecular principles was beginning to open new areas of research. Still, my personal introduction to these concepts remained theoretical. Several years after finishing my undergraduate studies, I had the opportunity to observe the embryonic development of a wide array of marine invertebrate species representing many different animal phyla at Friday Harbor Laboratories in the San Juan Islands, Washington. I watched as the cells of each animal began their program of development after fertilization—a program characteristic of each species—breathtaking in its complexity and beauty.

Later, while working in laboratories that employed fruit flies to understand the development of the nervous system and the establishment of the anterior-posterior axis in the early embryo, at Stanford University and Columbia University, respectively, I first appreciated the combined power of genetics and molecular biology. As a graduate student in the laboratory of Marian Carlson at Columbia, I used genetics and molecular biology to find and characterize genes in baker's yeast (another model organism) that regulate enzyme production in response to changes in sugar source. Finally equipped with the tools of genetics and molecular biology, I wanted to return to animal development for my post-doctoral research. I was attracted to elegant work on the *lin-12* gene and its role in development as conducted in the laboratory of Iva Greenwald at Columbia University. During the five and a half years I spent in Greenwald's laboratory (characterizing two genes functionally related to *lin-12*), I became increasingly enamored with the model organism on which my own research is now based, the worm *C. elegans*.

EGG-LAYING, CANCER, AND ALZHEIMER'S DISEASE

Although the story of *lin-12* is still unfolding, it provides an illustration of the model-organism approach and its unpredictable relevance to medicine. In the world of *C. elegans* genetics, *lin* is an abbreviation for "abnormal cell *lin*eage." Worms with abnormal *lin* gene activity exhibit very distinct changes in the lineage of certain cells during development. For instance, normally each worm has one passageway through which it lays eggs, a structure known as the vulva. The normal vulva consists of twenty-two cells that arise from the characteristic division pattern (lineage) of three vulval precursor cells during the early life of the worm. Defects in the vulval cell lineage can cause changes that ultimately block the passage of eggs out of the mother worm. These mutant animals are easy to recognize since they soon become full of eggs (which hatch inside the mother, eventually causing her destruction as the progeny escape). This easily detectable defect provided the basis for a forward genetic approach to the general problem of how a particular anatomical structure is formed—in this case, the vulva.[16] As is typical with the model-system approach, the genes defined by mutations in the *lin* genes are not likely to solve medical problems associated with human childbearing, but they have led to some unpredictable insights into human disease.

LIN-12 is a founding member of a conserved family of proteins now known to act as receptors for cell-cell communication.[17] Its role in development was discovered on the basis of animals bearing mutations in the *lin-12* gene: animals with mutations that either elevate or reduce *lin-12* activity can no longer lay eggs properly.[18] Moreover, elevation or reduction of *lin-12* activity has opposite effects on particular cell fate decisions. Greenwald characterized numerous *lin-12* mutations and cloned the *lin-12* gene.[19] The *lin-12* DNA sequence and subsequent experiments demonstrated that *lin-12* and its relatives in other organisms encode a cell-surface receptor protein. Activation of this receptor depends on interaction with a specific protein produced on a neighboring cell. During development, this receptor activity was found to govern many different binary cell fate decisions. That is, the activity of this receptor caused cells to "choose" one fate as opposed to an alternate fate that they would adopt in the absence of receptor activity. In and of itself, understanding mechanisms of development at the genetic and molecular level is fascinating to developmental biologists. But to those on the front lines of medicine, these results initially appeared to be of no apparent consequence. Until. . . .

Two years before I joined Greenwald's laboratory, Leif Ellisen and colleagues in the Jeffrey Sklar laboratory in the Department of Pathology, Brigham and Women's Hospital in Boston, discovered that a certain chromosome rearrangement associated with T lymphoblastic leukemia disrupts a gene very similar to the worm *lin-12* gene and the related *Notch* gene in fruit flies.[20] It was not clear, however, if this disruption would cause a reduction, increase, or change in gene activity. Since the genetic and molecular nature of *lin-12* and *Notch* were well known in worms and flies, workers could use these model organisms to, first, engineer a form of the worm and fly genes that resembled the disrupted human gene and, second, to reintroduce the gene into the model organism and see if the effect was consistent with a reduction, increase, or change in gene activity.[21] The answer was clear: an increase in activity of the receptor was caused by the truncation built to mimic the leukemia-inducing form of the gene. The cancer investigation could then proceed with a clear understanding of the nature of the effect of the mutation on the receptor. This result would have been very difficult to obtain using human tissue culture or other means. Since then, three additional *lin-12*/Notch-like genes have been found in humans, two of which have also been implicated in cancer. *Lin-12* mutations do not cause cancer in worms, and yet they are providing insights into the mechanisms of human cancer.[22] Furthermore, these genes have since been manipulated in mammals and are known to participate in many aspects of development, as they do in worms and flies.[23] Other genes that affect vulval development define additional pathways directly implicated in human cancers such the Ras-induced cancers and retinoblastoma.[24]

A more surprising result came several years later. When *lin-12* activity is elevated, the ventral surface of the worm has multiple protrusions (the so-called Multivulva phenotype), rather than the usual one small protrusion. One of my fellow postdoctoral researchers, Diane Levitan, wanted to find genes involved in the morphological changes affecting the ventral surface of the worm when *lin-12* activity is elevated. She looked for these genes using a genetic approach known as a suppressor screen. Starting with *lin-12* mutants that displayed the Multivulva phenotype, she mutagenized these animals and looked for mutations in other genes that would revert the anatomical defect to the normal morphology. This is an extraordinarily powerful approach widely used in many model systems to identify genes that are functionally related to a gene or process of interest. Much to everyone's surprise, the sequence of one of the genes defined by suppressor mutations, *sel-12*, turned out to be very similar to a gene that had just been identified in humans for its role in familial Alzheimer's disease.[25] Even more astonishing was the fact that the human gene, when introduced into the mutant worms, can function in the worm to reverse the vulval effects of a mutation that interferes with *sel-12* activity.[26] It has taken several more years of research in a number of different laboratories, but an answer to this unlikely connection is emerging. Alzheimer's disease is associated with the overproduction of a certain form of a protein in the brain known as ß amyloid. It turns out that very similar (if not identical) enzymes cause the formation of the aberrant ß amyloid protein and are important for the activation of LIN-12/Notch receptors.[27] One practical implication of the connection is that the pharmaceutical agents developed to treat and prevent Alzheimer's disease will have to take into account the possible unwanted effects on the human Notch receptors. Therefore, work that began by addressing the molecular basis for development is propelling forward two fields of direct relevance to human disease.

BOUNDLESS BIOMEDICAL RAMIFICATIONS

As the list of diseases associated with aberrant Notch pathway activity grows,[28] model-organism research continues to unravel the complexity of development, and unforeseen connections to human disease continue to emerge. In *C. elegans* alone, there are scores of equally interesting stories unfolding at a truly frantic pace. To give a glimpse of the remarkable ways in which this model organism is being used, I would like to briefly mention a few more of these stories.

Aging, Insulin, and Diabetes

Insulin is critical for the regulation of blood glucose levels; improper insulin receptor function is associated with diabetes. In worms the insulin receptor pathway controls an important developmental decision: whether to undergo reproductive development or to enter the dauer pathway. In addition, mutations in the worm insulin receptor gene confer longevity to worms. Follow-up studies have provided important molecular links between metabolism, aging, and disease.[29]

Reverse Genetics to Investigate Disease

The basis of many devastating heritable diseases is being discovered by studying genes of affected and unaffected families. In some cases, even though the disease-causing mutations have been identified, it is difficult to proceed from gene identification to an understanding of the biochemical and cellular consequences of the mutation—information that is crucial to design appropriate therapies. A reverse genetic strategy can be applied in model organisms such as *C. elegans*: the function of worm genes can be altered using molecular biology. Therefore the activity of a worm gene corresponding to a human so-called disease gene (a gene that when mutant in humans is known to be associated with disease) can be reduced or elevated in the worm and the consequences assessed.[30] One example employing this strategy is spinal muscular atrophy (SMA), which is caused by mutations in a gene well conserved in worms. Mutation of the worm gene causes several visible defects that can be exploited in further experiments. Extensions of this type of analysis in worms can include suppressor-genetics strategies to identify functionally related genes or molecules that may be targets for therapeutic agents.

The Physiological Effects of Drugs

Many drugs currently used to treat human disease have complex effects, only some of which can be found by chemical studies and drug trials. One example is drugs used to treat depression, some of which have effects on brain chemistry that are complex and not well understood. Worm neurons contain receptors and neurotransmitters similar to those in the human brain. One way *C. elegans* is being used to understand drug action is by defining the effects of drugs such as Fluoxetine (Prozac) on worm behavior and then looking for mutants resistant to these effects in order to identify genes that encode proteins affected by the drug. In this way, the worm is used to tell investigators what types of molecules the drug affects. So far, these studies confirm the effects of Fluoxetine on serotonin reuptake, but they also point to other potentially important effects that would be difficult to predict or infer in systems not amenable to genetic analysis.[31]

Bacterial Virulence Studies

Worm and bacteria researchers are using the genetics in both organisms to uncover the bacterial genes that promote virulence and the host genes that allow pathogenic access to the host.[32] For example, certain bacteria such as *Pseudomonas aeruginosa* are infectious to both *C. elegans* and to humans. Investigators have identified genes in the bacteria that are necessary for virulence in worms. There appear to be at least three different ways in which this bacterium can kill worms. Furthermore, workers identified a gene in worms that when mutant, renders the worms resistant to bacterial infection. The mammalian counterpart for this gene is expressed in tissues often infected by this bacteria.[33] Needless to say, an understanding of the human counterpart to these genes might open a new road for treating and preventing infections.

CONCLUDING REMARKS

It is my hope that my own laboratory at New York University will be the source of similar stories in the years to come. We study the *C. elegans* germ line (the tissue that gives rise to eggs and sperm). This tissue has always held a special fascination for me since it constitutes the link from generation to generation. Hence it is, in a way, immortal. Using a genetic approach, we have identified mutations that perturb the development of the germ line and are characterizing them. To me, the results are intriguing at many levels—from the simple standpoint of the development of the worm germ line to the general principles of development that might emerge, as well as the potential to discover a useful model system for human disease. So far, one of the genes we have "hit" is well characterized in both worms and humans, and our mutations are contributing new data to the field.[34] Another gene we identified is present in all eukaryotes, but it has not been previously studied in multicellular organisms.[35] We are optimistic that the identity and function of many other genes will be revealed through our genetic and molecular analyses.

NOTES

An earlier version of this article appeared as "Le potenzialità dei model organism," *KOS*, no. 176 (2000): 14–19.

1. The following Web sites contain information and links to model-organism research: for *C. elegans*, see the *Caenorhabditis elegans* WWW server (elegans .swmed.edu) and WormBase (www.wormbase.org). The National Institutes of Health (NIH) maintains a model organism site (www.nih.gov/science/models/

index.html) with links to major resources for all model organisms. The NIH also maintains PubMed (www.ncbi.nlm.nih.gov), a searchable database of the biomedical literature.

2. This information was gathered from links to the appropriate awards found at Nobelprize.org (www.nobel.se), a site containing detailed information on Nobel Prize awards. During the final copyediting of this chapter, two additional *C. elegans* researchers, Andrew Z. Fire and Craig C. Mello, were awarded the 2006 Nobel Prize in Physiology or Medicine for their discovery of RNA interference.

3. Regrettably, I have had to omit many primary references in the interest of space; where indicated, please refer to references within reviews cited in the notes below.

4. See Pamela M. Carroll et al., "Model Systems in Drug Discovery: Chemical Genetics Meets Genomics," *Pharmacology and Therapeutics* 99 (2003): 183–220.

5. Helpful background information can be found in any standard genetics textbook. Several such texts can be found online in the books link of the National Center for Biotechnology Information (NCBI) Web site (www.ncbi.nlm.nih.gov/ entrez/query.fcgi?db=Books). Gregor Mendel's results were published in Gregor Mendel, "Versuche über Pflanzenhybriden," *Verhandlungen des naturforschenden Vereines in Brünn, Bd. IV für das Jahr 1865*, Abhandlungen, 3–47; the nature of the genetic code was deduced by brilliant genetic experiments reported in F. Crick et al., "General Nature of the Genetic Code for Proteins," *Nature* 192 (1961): 1227–32.

6. Updated information on the genome sequences of model organisms can be found by following links on the NIH model organism Web site (www.nih.gov/science/ models/index.html).

7. Clifford W. Bogue, "Genetic Models in Applied Physiology: Functional Genomics in the Mouse; Powerful Techniques for Unraveling the Basis of Human Development and Disease," *Journal of Applied Physiology* 94 (2003): 2502–9.

8. Alfred Henry Sturtevant, *A History of Genetics* (1965; Cold Spring Harbor, N.Y.: Cold Spring Harbor Laboratory Press, 2001).

9. Maureen M. Barr, "Super Models," *Physiological Genomics* 13 (2003): 15–24.

10. Sydney Brenner, "The Genetics of *Caenorhabditis elegans*," *Genetics* 77 (1974): 71–94.

11. Sydney Brenner, "Letter to Max Perutz," quoted in *The Nematode Caenorhabditis elegans*, ed. W. B. Wood (Cold Spring Harbor, N.Y.: Cold Spring Harbor Laboratory Press, 1988), x–xi.

12. J. E. Sulston and H. R. Horvitz, "Post-embryonic Cell Lineages of the Nematode *Caenorhabditis elegans*," *Developmental Biology* 56 (1977): 110–56; J. E. Sulston, D. G. Albertson, and J. N. Thomson, "The *Caenorhabditis elegans* Male: Post-embryonic Development of Non-gonadal Structures," *Developmental Biology* 78 (1980): 542–76; J. E. Sulston et al., "The Embryonic Cell Lineage of the Nematode *Caenorhabditis elegans*," *Developmental Biology* 100 (1983): 64–119; Judith Kimble and David Hirsh, "The Post-embryonic Cell Lineages of the Hermaphrodite and

Male Gonads in *Caenorhabditis elegans*," *Developmental Biology* 70 (1979): 396–417.

13. The *C. elegans* Sequencing Consortium, "Genome Sequence of the Nematode *C. elegans*: A Platform for Investigating Biology," *Science* 282 (1998): 2012–18.

14. See WormBase for updated information on cloned genes (www.wormbase.org).

15. This information was taken from the Caenorhabditis Genetics Center at the University of Minnesota (biosci.umn.edu/CGC/CGChomepage.htm) accessed Feb. 2, 2004.

16. H. Robert Horvitz and John E. Sulston, "Isolation and Genetic Characterization of Cell-Lineage Mutants of the Nematode *Caenorhabditis elegans*," *Genetics* 96 (1980): 435–54; John E. Sulston and H. Robert Horvitz, "Abnormal Cell Lineages in Mutants of the Nematode *Caenorhabditis elegans*," *Developmental Biology* 82 (1981): 41–55; Iva S. Greenwald, Paul W. Sternberg, and H. Robert Horvitz, "The *lin-12* Locus Specifies Cell Fates in *Caenorhabditis elegans*," *Cell* 34 (1983): 435–44; Edwin L. Ferguson and H. Robert Horvitz, "Identification and Characterization of Twenty-Two Genes that Affect the Vulval Cell Lineages of the Nematode *Caenorhabditis elegans*," *Genetics* 110 (1985): 17–72.

17. Many reviews on *lin-12*/Notch signaling are available including Iva Greenwald and Gerald M. Rubin, "Making a Difference: The Role of Cell-Cell Interactions in Establishing Separate Identities for Equivalent Cells," *Cell* 68 (1992): 271–81; Iva Greenwald, "LIN-12/Notch Signaling: Lessons from Worms and Flies," *Genes and Development* 12 (1998): 1751–62; Sarah Bray, "Notch Signalling in *Drosophila*: Three Ways to Use a Pathway," *Seminars in Cell and Developmental Biology* 9 (1998): 591–97; Spyros Artavanis-Tsakonas, Matthew D. Rand, and Robert J. Lake, "Notch Signaling: Cell Fate Control and Signal Integration in Development," *Science* 284 (1999): 770–76; Jeffrey S. Mumm and Raphael Kopan, "Notch Signaling: From the Outside In," *Developmental Biology* 228 (2000): 151–65; Martin Baron et al., "Multiple Levels of Notch Signal Regulation," *Molecular Membrane Biology* 19 (2002): 27–38; M. Baron, "An Overview of the Notch Signalling Pathway," *Seminars in Cell and Developmental Biology* 14 (2003): 113–19.

18. Greenwald, Sternberg, and Horvitz, "The *lin-12* Locus Specifies Cell Fates."

19. Ibid.; Iva Greenwald, "*lin-12*, a Nematode Homeotic Gene, Is Homologous to a Set of Mammalian Proteins that Includes Epidermal Growth Factor," *Cell* 43 (1985): 583–90; John Yochem, Kathleen Weston, and Iva Greenwald, "The *Caenorhabditis elegans lin-12* Gene Encodes a Transmembrane Protein with Overall Similarity to *Drosophila* Notch," *Nature* 335 (1988): 547–50.

20. Leif W. Ellisen et al., "*TAN-1*, the Human Homolog of the *Drosophila Notch* Gene, Is Broken by Chromosomal Translocations in T Lymphoblastic Neoplasms," *Cell* 66 (1991): 649–61.

21. Gary Struhl, Kevin Fitzgerald, and Iva Greenwald, "Intrinsic Activity of the Lin-12 and Notch Intracellular Domains *In Vivo*," *Cell* 74 (1993): 331–45.

22. Ivan Maillard and Warren S. Pear, "Notch and Cancer: Best to Avoid the Ups and Downs," *Cancer Cell* 3 (2003): 203–5; Brian J. Nickoloff, Barbara A. Osborne,

and Lucio Miele, "Notch Signaling as a Therapeutic Target in Cancer: A New Approach to the Development of Cell Fate Modifying Agents," *Oncogene* 22 (2003): 6598–608.

23. Thomas Gridley, "Notch Signaling in Vertebrate Development and Disease," *Molecular and Cellular Neuroscience* 9 (1997): 103–8.

24. David S. Fay and Min Han, "The Synthetic Multivulval Genes of *C. elegans*: Functional Redundancy, Ras-Antagonism, and Cell Fate Determination," *Genesis* 26 (2000): 279–84.

25. Diane Levitan and Iva Greenwald, "Facilitation of *lin-12*–Mediated Signalling by *sel-12*, a *Caenorhabditis elegans* S182 Alzheimer's Disease Gene," *Nature* 377 (1995): 351–54.

26. Diane Levitan et al., "Assessment of Normal and Mutant Human Presenilin Function in *Caenorhabditis elegans*," *Proceedings of the National Academy of Sciences* USA 93 (1996): 14940–44.

27. Dennis Selkoe and Raphael Kopan, "Notch and Presenilin: Regulated Intramembrane Proteolysis Links Development and Degeneration," *Annual Review of Neuroscience* 26 (2003): 565–97.

28. Thomas Gridley, "Notch Signaling and Inherited Disease Syndromes," *Human Molecular Genetics* 12 (2003) Review Issue no. 1: R9–R13.

29. Donald W. Nelson and Richard W. Padgett, "Insulin Worms Its Way into the Spotlight," *Genes and Development* 17 (2003): 813–18.

30. Emmanuel Culetto and David B. Sattelle, "A Role for *Caenorhabditis elegans* in Understanding the Function and Interactions of Human Disease Genes," *Human Molecular Genetics* 9 (2000): 869–77.

31. Robert K. M. Choy and James H. Thomas, "Fluoxetine-Resistant Mutants in *C. elegans* Define a Novel Family of Transmembrane Proteins," *Molecular Cell* 4 (1999): 143–52; William R. Schafer, "How Do Antidepressants Work? Prospects for Genetic Analysis of Drug Mechanisms," *Cell* 98 (1999): 551–54.

32. Rosanna A. Alegado et al., "Characterization of Mediators of Microbial Virulence and Innate Immunity Using the *Caenorhabditis elegans* Host-Pathogen Model," *Cellular Microbiology* 5 (2003): 435–44.

33. Man-Wah Tan et al., "*Pseudomonas aeruginosa* Killing of *Caenorhabditis elegans* Used to Identify *P. aeruginosa* Virulence Factors," *Proceedings of the National Academy of Sciences* USA 96 (1999): 2408–13.

34. Anita S.-R. Pepper, Darrell J. Killian, and E. Jane Albert Hubbard, "Genetic Analysis of *Caenorhabditis elegans* glp-1 Mutants Suggests Receptor Interaction or Competition," *Genetics* 163 (2003): 115–32; Anita S.-R. Pepper et al., "The Establishment of *Caenorhabditis elegans* Germline Pattern Is Controlled by Overlapping Proximal and Distal Somatic Gonad Signals," *Developmental Biology* 259 (2003): 336–50.

35. Darrell J. Killian and E. Jane Albert Hubbard, "*C. elegans pro-1* Activity Is Required for Soma/Germline Interactions that Influence Proliferation and Differentiation in the Germ Line," *Development* 131 (2004): 1267–78.

The Troop Trope: Baboon Behavior as a Model System in the Postwar Period

SUSAN SPERLING

In order to get to monkeys, I need to begin with Clifford Geertz. According to Geertz, we are the animals suspended in self-spun "webs of significance." Geertz has construed culture to be the webs and has suggested that anthropology is less an experimental science than an interpretative one. Geertz's reflections appeared in *The Interpretation of Cultures*, during a period in the discipline's history in which the seeds of poststructuralist interpretation were beginning to sprout in anthropology programs largely influenced by various permutations of structural functionalism.[1] Structural functionalism, British social anthropology's key contribution to twentieth-century social science, explains the structural pattern of social institutions in terms of how they function as integrated systems to fulfill individual and societal needs.[2]

The anthropologist Micaela di Leonardo refers to this tradition, and its many offshoots, as "the science and order" corner of anthropology, in which, by the 1970s, had gathered a number of strange bedfellows:[3] "Levi-Straussian structuralism, originating in studies of linguistics and folklore, had percolated outward across disciplines and was taken up and fused with Marxism . . . structuralist Marxism promised to set Marxist analysis once again upon a scientific footing to allow the clear taxonomy of societies."[4] Structuralism strongly influenced anthropology in the study of symbolic systems, in the work of a number of anthropologists such as Geertz in the early postwar period and the later feminist analysis of Sherry Ortner.[5] A unifying theme of both structural functionalism and structuralism is the assertion of widespread social and cultural regularities across time and space.

By the 1970s, some radical turns had reframed aspects of the discipline as anthropologists reenvisioned the field vis-à-vis its relationship to colonization. In this vein, Talal Asad and others investigated the phenomenology of ethnography. This historical and reflexive trend led to works like Eric Wolf's *Europe and the People without History*, in which third and fourth world peoples were studied within the context of European colonization and global capitalism.[6]

A disciplinary tree with its roots in Victorian evolutionism had grown many generations of new wood. Structural functionalism and structuralism, with their emphasis on ethnographic data collection and the synchronic study of institutions, had reformulated the discipline as empirical science. By the 1980s, the leafy branches were increasingly reflexive, literary, iconoclastic, and antirationalist.

To return to Geertz: if we are animals suspended in webs that we ourselves have spun, what, then, is a baboon? The baboon and other nonhuman primate troops ranging across the savannas of postwar anthropology at the historical moment of this reflexive turn seem an amalgam of the primitive human progenitors of the Victorian imagination, along with the third world Others of structural-functionalist anthropology.[7] They are not suspended in their own webs of meaning, yet they have complex social relations that can be subjected to structural-functionalist analysis by the objective human observer. For many evolutionary anthropologists, the baboon troop, as a model system for the analysis of human cultural origins, would help answer some of anthropology's oldest and most fundamental questions about the meanings of human behavior.

The baboon troop's appearance as a legitimate object of ethnographic/ethological study is thus not accidentally coincident with the loss of "science and order" in the socio-cultural branches of the discipline. The evolutionary analysis of primate behavior is where science and order found a perch in an otherwise largely hostile discipline. By 1975, the cultural anthropologist Robin Fox was explicit about his use of nonhuman primates as replacements for so-called human primitives:

> Older theorists speculated on the "earliest conditions of man," and we know debates raged between proponents of "primitive promiscuity" and "primitive monogamy." The former was usually seen as a prelude to "matriarchy" (now popular again) and the latter to "patriarchy." This has been dismissed for well-known reasons. But I think we can now go back to the question in a different way. We know a great deal about primates which can tell us what is behaviorally available to our order in general and, therefore, what must have been available by way of a behavioral repertoire to our ancestors . . . "early man" then, in this sense, was less like modern man gone wild than like a primate tamed. *And even if we cannot deduce accurately kinship systems of early man from those of the most primitive humans, we can do something better, we can distill the essence of kinship systems on the basis of comparative knowledge and find the elements of such systems that are logically, and hence in all probability chronologically, the "elementary forms of kinship."*[8]

Here, Fox traced the evolution of human kinship through the primates, borrowing, as he admits, "somewhat recklessly from the jargon of social anthropology, descent and alliance." He goes on to say: "The real question is do the (human) rules represent more than a 'labeling' procedure for behavior that would occur anyway? If group A and B were called 'Eaglehawk' and 'Crow,' and the various lineages 'snake,' 'beaver,' 'bear,' and 'antelope,' etc., then a picture emerges of a proto-society on a clan moiety basis."[9] Examples of this positioning of the baboon troop in the cosmological slot for "primitive society" abound in the literature of the period. Nonhuman primates, particularly savanna baboons, became the missing link in the evolutionary models of the late 1960s and 1970s.

But focusing only on theoretical shifts within anthropology that replaced human primitives with nonhuman primates in evolutionary models is to view primatology through too narrow a lens. Other trends proved important in the baboonization (the primatologist Linda Fedigan's term) of human evolution and culture, from low-cost airfare to game reserves in which these large cercopithecine monkey were fairly easy to observe to the widespread availability of antibiotics. The intertwining of these material factors with the shift away from empiricism within mainstream cultural anthropology positioned nonhuman primates as important behavioral models for primitive human society within a discipline that would seem to have been well and done with the very concept of "the primitive."

WHAT IS A BABOON?

Baboons (genus *Papio*) are large monkeys living in mixed-sex troops widely distributed over the African continent. There are great similarities in the social behavior of the twenty-five to thirty species of the order Cercopithecinae, to which the baboons belong.[10]

Cercopithecines have proven very successful in evolutionary terms. They are abundant throughout many habitats in Africa and Asia. Baboons occupy a wide range of habitats throughout sub-Saharan Africa including rain forest, woodland, coastal, and high-altitude areas and semidesert savanna. Savanna baboons are generalized feeders eating grass, bulbs, rhizomes, flowers, fruits, leaves, tuber seeds, and tree gum. Adult males, and other animals to some extent, supplement this diet through the opportunistic hunting of a variety of mammals, including monkeys of other species.[11]

Food is unevenly distributed in the savanna environments, and baboons, particularly males, are large. Troops range across a large area to feed. In Amboseli (Kenya), an arid savanna, troops travel about 5.9 km a day, and home ranges (the area traveled over an extensive period) are about 24 km. Savanna population densities are apparently low where animals depend mostly on roots and grasses, while in forested areas, where baboons are more frugivorous, they have smaller home ranges and higher population densities.[12]

There are consistent cercopithecine patterns of social structure and behavior. Males typically migrate out of their natal groups when they reach sexual maturity and often several more times over their life spans. Females are philopatric, staying in their natal groups all their lives. This stands in contrast to chimpanzees, in which young females leave their birth group. As Don J. Melnick and Mary C. Pearl point out, this dispersal pattern, presumably a very old one in cercopithecine evolutionary history, shapes aspects of the genetic and social structure of populations.[13] Baboons are quite sexually dimorphic, with females weighing around fourteen kilograms and males twenty-five kilograms. They are quadrupeds, sleeping in trees or on rocky outcroppings during the night and coming down to the ground to feed during the day.

PRIMATE PATTERNS

Primates vary considerably in size, structure, and behavior. Studies of the fossil record and molecular evolutionary data indicate that primates descended from early mammalian ancestors present during the adaptive radiation of the placental mammals about 65 million years ago. Approximately 70 percent of all living primates (about 130 species) are monkeys. The first clearly identifiable primate fossils are found in the Eocene (53–57 mya) and resemble living lorises and lemurs (prosimians). During this period more than sixty prosimian-like genera have been identified. By the end of the Oligocene (37–22.5 mya), the fossil record suggests that Old World monkeys and hominoids were evolving along separate evolutionary pathways. This area of primate paleontology remains very ambiguous. The richest data come from the Fayum (Egypt) between 35 and 33 million years ago and demonstrate anthropoid radiation along a number of lineages.[14]

The early primates shared certain characteristics still present in the order, including humans (i.e., primitive primate characteristics). This list includes: (1) the retention of five digits; (2) nails instead of claws; (3) flexible hands and feet; (4) a tendency toward erectness; (5) the retention of the clavicle; (6) a generalized dental pattern and lack of specialization in diet; (7) a reduction of the snout; (8) increased emphasis on vision; (9) an expansion and increased complexity of the brain; (10) long gestation for size; (11) dependence on flexible and learned behavior; and (12) adult males often associating permanently with the group.[15]

Modern baboons and humans also have a variety of evolutionary-derived characteristics—that is, specialized features evolved over the two species' separate evolutionary histories. These include in baboons great sexual dimorphism (adult males are twice the size of adult females) and large, razor-sharp incisors, the roots of which extend along the snout, giving the baboon its characteristic doglike face. Female baboons also develop a large and vivid estrus swelling of the perineal area around the period of ovulation. The migration of males out of groups is so widespread among genetically distinct cercopithecines living in different habitats that is hard not to conclude that some adaptive pressures have at least partially shaped this pattern.

Derived characteristics of recent evolutionary origin in our species include an increasing dependence on bipedalism, a relatively enormous cerebral cortex with an increased neural complexity of areas of the motor cortex and centers of communication, the generation of symbolically mediated communications, and the production of human culture.

THE TROOP TROPE: BABOON BEHAVIOR AS A MODEL SYSTEM

Modern Western field studies of primates were first undertaken in decolonized Africa and other third world sites in the period following World War II. Soon thereafter, anthropological primatologists and their advocates in other disciplines began to fit data about monkeys and apes into models of human evolution. The template for this enterprise had been set early in the century by Robert M. Yerkes and Clarence Ray Carpenter, primatologists whose work Donna Haraway has extensively examined,[16] but it is in the postwar period that primatology flourished. Field and laboratory observations of primates have produced a large corpus of data on the behavior of diverse species. The integration of these facts into models of human evolution has consumed over three decades of scholarship.

When I began my studies as a graduate in physical anthropology at Berkeley in the 1970s, modern primate studies had emerged from a period in which a relatively small number of researchers collected natural histories of various primate species and had entered an era of widespread structural-functionalist model building. The first period, which might be called the natural history stage of primate studies, occurred roughly between 1950 and 1965.[17] In the second stage (from the mid-1960s to the mid-1970s), data from a variety of field studies, particularly those of savanna baboons and chimpanzees at the Gombe reserve in Tanzania, were incorporated into structural-functionalist frameworks for the evolutions of human society. These studies focused on the sexual division of labor, the origins of the family, and the origins of human gendered behavior. The third phase began in the 1970s with the advent of sociobiology as the functionalist model par excellence for understanding primate evolution.[18]

During the second stage of modern primatology, structural-functionalist analysis became central to the problem-oriented studies that replaced the earlier emphasis on natural history. Primatology has always been a heterogeneous field spread across at least three different disciplines that includes research on proximate casual factors affecting social behavior and on the complex interaction between social structure, behavior, and ecology (socioecology). But it is the structural-functionalist grand theory builders, those who focus exclusively on ultimate causality, who have been the progenitors of the most influential and widely popularized visions of primate behavior. In these models of the 1960s and 1970s, all aspects of troop structure were explained as adaptive mechanisms, selected during a species history because they "functioned" to promote survival.[19]

Linda Fedigan points out that the second-stage baboonization of early human life was based on a savanna environment hypothesis for early hominid evolution (the fossil evidence for this is now equivocal).[20] Presumably, early hominids would have shared certain selective pressures with modern baboon troops, particularly for predator protection by large males. Sherwood Washburn, Irven DeVore, and other early baboon researchers viewed male dominance as functioning to organize troop members hierarchically and to control overt aggression.[21] Fedigan has argued that the other primary model for protohominid evolution, the chimpanzee, appears preferable. Here the analogy is based on a phylogenetic relationship between chimp and human immensely closer than between either species and baboons. This model initially emphasized the mother-offspring bond and sharing with the matrifocal family. The first wave of feminist interpreters of primate behavior, particularly the constructors of the "woman the gatherer" model,[22] would turn here for evidence of the centrality of females in early hominid evolution.

The second-stage savanna baboon model tended to support a Hobbesian view of human society, while the chimpanzee model originally reflected a more benign vision, stressing the mother-infant pair and a flexible, less hierarchical social structure. But in each case, monkey and ape groupings were viewed as model systems for studying human social behavior and political life. The baboonization of protohominids became so common that by the mid-1970s not a single introductory text in human evolution omitted reference to it. As the zoologist Thelma Rowell and other critics of the model stressed,[23] many of the generalizations about the functions of male dominance made during this period were unsubstantiated by data from increasingly diverse researchers and research sites. Rowell's studies of troop movements among the forest baboons, for instance, indicated that the direction of daily foraging routes was determined by a core of mature females rather than by dominant males. As Donna Haraway and Sandra Harding have noted, women primatologists have often had a different vision of a group's structure and behavior because they attended to female actors in ways ignored by male primatologists.[24] This focus on female behavior in baboons and other species fueled critical deconstructions of the baboon model of the 1970s. A number of studies questioned the central assertion that male dominance conferred a reproductive advantage, thus contributing to selection for male aggression.

A 1978 article that appeared in *Psychology Today* illustrates the diffusion of the image of the baboon troop as a model system into popular media:

> In all primate societies, the division of labor by gender creates a highly stable social system, the dominant males controlling territorial boundaries and maintaining order among lesser males by containing and preventing their aggression, the females tending the young and forming alliances with other females. Human primates follow this pattern so remarkably that it is not difficult to argue for biological bases for the type of social order that channels aggression to guard the territory which in turn maintains an equable environment for the young.[25]

Although the template here is the savanna baboon troop as described by Washburn and DeVore—and contested early in this period by Rowell and others who called into question many of the assumptions of the savanna model—"all primate species" collapsed the diversity and specificity of data primates into a single category, "primate societies."

The ubiquitous baboon troop now firmly occupied the position of primitive society in the schemas of nineteenth-century anthropologists. The diffusion of cultural relativism into modern social science had made it untenable to fit extant human communities into this early evolutionary slot. Baboons and other nonhuman primates became the early ancestral group from which human institutions were seen to have evolved. The substitution of primates for human primitives thus neatly retained an important Western cosmological category for use in the era of decolonization, the discrediting of prewar eugenics, and the construction of the third world.

One of the consequences of this key insertion of the nonhuman primate into the Western symbolic niche for primitive progenitor was an implied obliteration of the border between human and nonhuman. What is meant by such terms as the *division of labor* when referring to baboons? Does this term mean the same thing when applied to human groups? Monkeys and apes do not have a division of labor along gender lines as do human cultures; each animal performs subsistence tasks in approximately the same way as the others, consuming on the spot what is individually foraged. Human divisions of labor by sex are complex historical and socioeconomic phenomena embroidered with symbolic meanings that seem unavailable to animals.

THE ASCENT OF *HOMO GENETICUS*

The 1975 publication of *Sociobiology: The New Synthesis* by the Harvard entomologist Edward O. Wilson constituted a signal event across the disciplines concerned with animal and human behavior and evolution.[26] In it, Wilson made two major assertions: that all important social behaviors are genetically controlled and that the natural selection of the genome is caused by a set of specific adaptive mechanisms (kin selection) that produce behaviors maximizing an organism's ability to contribute the greatest number of genes to the next generation. The historical roots of sociobiology lie in nineteenth- and early twentieth-century arguments about the level at which natural selection operates, that of the group or that of the individual. Evolutionists like Charles Darwin, John B. S. Haldane, and Vero C. Wynne-Edwards contended that traits may be selected because they prove advantageous for populations. In the 1960s and 1970s, William Hamilton and Robert Trivers proposed that traits can be selected only at the individual level and that all social behaviors are tightly controlled genetically.[27]

Hamilton's theory of kin selection is based on the idea that an organism's fitness has two components: fitness gained through the replication of its own genetic material by reproduction and so-called inclusive fitness gained from the replication of its own genes carried by relatives as a result of its actions. According to this theory, when an organism behaves altruistically toward related individuals, fitness benefits to kin also benefit the organism, but the actor's benefits are devalued by the coefficient of relatedness between actor and relatives. Thus genes are viewed as being selected because they contribute to their own perpetuation, regardless of the organism of which they form a part. Trivers defined reciprocal altruism as behavior that appears to be altruistic but that, given the mutual dependence in a group, may be selected if it confers benefits on the altruist.

In *Sociobiology*, Wilson suggested that the social and biological sciences be subsumed by sociobiology. Not surprisingly, many viewed the idea of their discipline's somewhat cannibalistic incorporation into the body of sociobiology as an unsavory prospect. Some objected on political grounds to its explicit reductionism, viewing sociobiology as a revival of social Darwinism. The Boston-based collective Science for the People issued a blistering critique along these lines in 1976.[28] At the same time, sociobiology established a foothold in American and European departments of anthropology, zoology, and psychology. The American Anthropological Association sponsored a two-day symposium on sociobiology at its 1976 yearly meeting, and by the late 1980s, sociobiology had become the dominant paradigm among anthropological primatologists, replacing the structural-functionalist models of the second period of primatology.

There are many reasons for the ascendance of sociobiology over structural functionalism in primate studies. With the insertion of genes, sociobiologists claimed the terrain of biological science, thus attempting to resolve some of the ambiguities of stage-two studies through the application of a new theoretical model with testable hypotheses. Sociobiology thus shifted the explanatory model from science to biology. The sociobiological lexicon also incorporated terms from economics, most important among them, game theory. Now primate behavior would be described in terms of evolutionary trade-offs and strategies. At the same time, a kind of *épatez les féministes* (let's impress the feminists) made its appearance in descriptions of primate behavior with the inclusion of terms like *rape, coyness,* and *infanticide.*

BABOONS WITH BRIEFCASES

An Asian colobine, the Indian gray langur (*Presbytis entellus*), which I studied in graduate school, figured prominently in the sociobiological literature, largely through the work of Sarah Hrdy, a student of Irven DeVore's at Harvard.[29] Seeming anomalies in langur behavior, such as the caregiving to infants by females other than the mother, were now explained as reproductive strategies. As I have argued elsewhere,[30] feminist revisionism within primatology heavily incorporated the genetic game theory of sociobiology. The long-term studies of Jeanne Altmann and Shirley Strum also reflect this shift.[31] In the work of both scientists, the analysis of female behaviors and female-male interactions proves central, moving away from the androcentric narratives of researchers from stages 1 and 2 such as Washburn and DeVore with their sole focus on male hierarchies and on forms of cooperation in maintaining troop order.[32]

In her 1980s publication on the "pumphouse baboons" of Gilgil, Kenya, Shirley Strum described "strategically reciprocating" male and female social relations in which males gained reproductive advantage by aligning with females. Strum asserted that a male's network of female relationships had more to do with his reproductive success than his position in the agonistic male hierarchy. In these studies, female choice was seen as a key factor in determining male reproductive success.[33]

Altmann, Strum, and others described baboon society during this period as "matrilineal," with kinship the key factor in organizing the group. Females were observed to perform many of the roles earlier attributed to males (e.g., policing, protecting, and leading). Females were described as having a stable dominance hierarchy, not the fluid relationships based on association with males limned by DeVore, Washburn, and Hall. Strum concluded during this period that males, in fact, had a rather fluid and unstable dominance hierarchy and that access to limited resources did not depend on male rank. In Altmann's studies of the 1980s at Amboseli, baboon females are described explicitly as "duel career" mothers, providing for their young while carefully sizing up and acting on affiliations that provide reproductive advantage in the context of their greater knowledge of troop relations over time and troop movement patterns in the home range. Now everyone ate power lunches on the savanna!

The sociobiological view of female primates as competitors attracted much public interest, as reflected in its widespread dissemination through the popular journalism of the period. Here, human females were portrayed as bearers of behavioral homologies from baboons and other nonhuman primate ancestors. DeVore, prominent during the second stage of baboon studies, adopted the new theory and lexicon of sociobiology. For example, he interpreted soap operas in terms of ubiquitous primate female reproductive strategies:

> Soap operas have a huge following among college students, and the female-female Competition is blatant. The women on these shows use every single feminine wile. On the internationally popular soap *Dynasty*, for example, a divorcee sees her ex-husband's new wife riding a horse nearby. She knows the woman to be newly pregnant so she shoots off a gun—which spooks the horse, which throws the young wife, and makes her miscarry.
>
> . . . Whole industries turning out everything from lipstick to perfume to designer jeans are based on the existence of female competition. The

business of courting and mating is after all, a negotiation process, in which each member of the pair is negotiating with those of the opposite sex to get the best deal possible, and to beat out the competition from one's own sex.[34]

NEW EVIDENCE OF PRIMATE DIVERSITY

During the 1980s, studies of relatively unknown species proliferated. A number of factors influenced this trend, including increased graduate enrollments and newly accessible sites. The arboreal monkeys of the Americas, as well as those of the Old World, had always proven extremely difficult to study because social behavior carried out in the trees is hard to follow and record. New data-collection techniques and perspectives from ecology began to have an important influence. Some primate researchers turned to sociobiology in an attempt to understand variation in group structure and behavior among arboreal species in the New World. Concepts and methodologies from ecology, such as optimal foraging theory, shifted the focus in many studies toward hypotheses about how animals interacted with their habitats. Socioecologists within primatology now measured the nutritional content of foods and studied feces to obtain specific data on consumption in order to correlate social structure with habitat use, emphasizing proximate factors in primate environments. A variety of biological samples became amenable to analysis during this period. Researchers took blood in order to study genetic patterns of relatedness within troops and populations as a whole. These data were often linked to sociobiological cost-benefit analyses.

Studies of primate cognition during this period revealed some of the complexity of primate intelligence, including the ability of apes in captivity to learn simple symbolic associations (ape languages). Longitudinal field studies of chimpanzees showed a very different kind of social organization from that of the cercopithecines. Here, related males formed the stable core of the group, with females migrating out of the loosely knit local populations at adolescence. The former image of the peaceful and idyllic chimp society faded as seemingly ubiquitous acts of violence, including infanticide and cannibalism, were recorded at the Gombe and Mahale sites. It also became clear that populations of chimpanzees in different ecological settings adopted different food-getting patterns.[35]

Fedigan and Strum characterize the present state of primatology (post stage 3) as one of "fragmentation and specialization" that has rejected the extreme reductionism of early sociobiology. Reflecting on this new attention to complexity, they write: "Today, baboons everywhere have more options in our images of their societies than ever before."[36] I would put it differently. While many recent studies have moved away from the strict reductionism of early sociobiology, a major shift in theoretical emphasis has not occurred. Reproductive strategies are still understood by most primatologists as the sine qua non of evolution. This approach in some ways lies outside the mainstream of modern biological science's focus on plasticity and contingency in the development of organisms. Sociobiology continues to diffuse into the arena of human ethology under the rubric of evolutionary psychology. At the end of the past millennium and at the beginning of the next, images of the baboon troop reflect what might be called the expanded sociobiological frame. Proximate factors affecting an individual or troop (patterns of mitigation, survivorship, foraging strategies, and other socioecological events and linkages) are now understood to contribute to individual fitness.

PRIMATOLOGISTS IN THE EPISTEMOLOGICAL JUNGLE

The baboon troop became a preeminent model system in postwar studies of human evolution as a result of complex events within anthropology and a number of Western cultures, particularly the United States. Access to field sites, antibiotics, cheaper airfares, and the ubiquitousness of baboons made them a seemingly natural subject for study. Accounts of baboon behavior over the past decades have reflected ideological influences on American culture, from second-wave feminism to the resurgence of social Darwinism in the work of sociobiologists and evolutionary psychologists. As Angela Creager points out in this volume, model systems are subject to "the recalcitrance of nature," and the baboon troop turns out to be a more complex and contingent grouping of individuals with varying behaviors and histories than many early studies assumed.

As exemplars of the early stages of human evolution, are baboons "the right tool for the job?"[37] Fossil and genetic evidence support a very recent time of divergence between the African apes and hominids. The savanna analogy that justified the baboon troop as a model system for understanding early human life has lost popularity as early fossil hominids are recovered from a variety of ancient habitats, including forests. Longitudinal studies of chimpanzees and of the closely related bonobos have offered model systems of social interaction in more complex and neurologically, anatomically, and genetically closer relatives. Even so, chimpanzees and bonobos may also not be "the right tool for the job" for modeling the origins of human culture.

A half century of natural history studies of nonhuman primates has revealed some of the relatively conservative aspects of neuroanatomy, reproductive physiology, social structure, and behavior across the order. Some of these
patterns are so ubiquitous among the over two hundred species in the order
(including humans) that they are reasonably hypothesized to be homologies
inherited from ancestral populations that arose about 60 million years ago during the adaptive radiation of many mammalian species including the primates.
As baboons or chimpanzees are studied as model organisms for evolutionarily
conservative systems, much heuristic data have been collected. For instance,
knowledge of the flexibility of primate female reproductive phenomena (in
response to social and other environmental factors) has generated important
hypotheses and data collection about nonhuman and human reproductive patterns.

But have we learned much of anything about human *cultural* patterns that
are of recent origin and mediated by our highly derived neurological structures—structures we do not share with baboons, chimpanzees, or bonobos?
Chimpanzees and bonobos have the ability to make simple tools and very
simple symbolic associations, but neither has human culture. The more derived
a structure or behavior, the less amenable to study it is through the use of a
different, but related, organism as a model system. Such is the case with the
political, economic, religious, and social patterns of human life. While the behavior of nonhuman primates can illuminate aspects of our common ancestry,
cultural patterns are not part of our common heritage with other primates.

The farther we move from the study of the primitive primate adaptive complexes, the more equivocal the application of monkey or ape data to humans
becomes, and the greater the tendency among the researchers to anthropomor-
phize animals. Baboons, chimps, and humans have quite distinct evolutionary histories, with many derived characteristics evolving over a long period of
separation from the common ancestor. The estrus swelling of females and the
razor-sharp canines of both males and females are part of a complex of *Papio*
reproductive structures and behaviors quite different from anything found in
humans.

In Geertz's essay in this volume, he examines the anthropological use of
ritual as a model system and reminds us of the anthropologist Victor Turner's
analysis of passage ceremonies in their transformative function—to "indicate
and constitute" transitions between health and sickness, childhood and adulthood, and life and death: "It is in the movement of people, or more exactly of
groups of people, through the dense allegories of ritual, 'the forest of symbols,'
that human community is formed, reformed, and held in place." The neuro-

anatomy underlying such a symbolic transformation appears to have developed quite recently in human evolution, even if, as Geertz indicates, the "forest of symbols" has been "thinned or cut down" in contemporary life.[38] (Palentological/archaelogical evidence suggests this change occurred between 1.8 million and one hundred thousand years ago, long after the divergence from our common ancestor with either baboons or chimpanzees.) For baboons to work as models for human cultural origins, one must assume that culture is merely an overlay or elaboration on baboon patterns of association. That is, at the very least, debatable. If primate studies are to be more than Darwinian fundamentalism, then primatologists must engage these epistemological questions more closely.[39]

We still know relatively little about how the specialized structures of the human neocortex create culture. To return again to Geertz in this volume, citing Bronislaw Malinowski on the function of magic ritual: we are often led to an impasse when gaps in our knowledge and the limitations of the human powers of observation betray us "at a crucial moment." The trope of the baboon troop as a model system for human origins has filled such a crucial gap in late-twentieth-century knowledge. Thus are scientific rituals engendered.

NOTES

This paper was originally presented to the History of Science Workshop "Model Systems, Cases, and Exemplary Narratives," organized by Angela Creager, Elizabeth Lunbeck, and Norton Wise at Princeton University in 2000. I particularly wish to thank Angela Creager for her reflections and suggestions, as well as the other workshop participants for their unique critique of these ideas. Some of this chapter's section on the influence of structural-functionalism in primate studies appeared first in *Gender at the Crossroads of Knowledge: Feminist Anthropology in the Postmodern Era*, edited by Micaela di Lenoardo (Berkeley: University of California Press, 1991).

1. Clifford Geertz, *The Interpretation of Cultures: Selected Essays* (New York: Basic Books, 1973).

2. Henrika Kucklick, *The Savage Within: The Social History of British Anthropology* (Cambridge: Cambridge University Press, 1991).

3. Micaela di Leonardo's introduction in *Gender at the Crossroads of Knowledge: Feminist Anthropology in the Postmodern Era* (Berkeley: University of California Press, 1991) offers an important critical review of these theoretical turns, especially as they were reflected in gender studies.

4. Ibid., 19.

5. Sherry Ortner, "Is Female to Male as Nature Is to Culture?" *Feminist Studies* 1

(1972): 5–31. The essay was reprinted in Michelle Zimbalist Rosaldo and Louise Lamphere, eds., *Woman, Culture, and Society* (Stanford: Stanford University Press, 1974), 7–88.

6. Talal Asad, ed., *Anthropology and the Colonial Encounter* (London: Ithaca Press, 1973); Eric Wolf, *Europe and the People without History* (Berkeley: University of California Press, 1982). As di Leonardo points out in *Gender at the Crossroads of Knowledge*, in the 1970s and 1980s, structuralism and structural functionalism declined as the iconoclastic visions of poststructuralism rose in large part from literary roots. Poststructuralism's central claim is the primacy of texts (scientific reports, narrative history, novels, advertisements, etc.) over observed behaviors in the analysis of culture.

7. Donna Haraway, *Primate Visions: Gender, Race, and Nature in the World of Modern Science* (Routledge: New York, 1989); and Susan Sperling, "Baboons with Briefcases vs. Langurs in Lipstick: Feminism and Functionalism in Primate Studies," in *Gender at the Crossroads of Knowledge: Feminist Anthropology in the Postmodern Era*, ed. Micaela di Leonardo (Berkeley, University of California Press, 1991), 204–34.

8. Robin Fox, ed., *Biosocial Anthropology* (New York: Wiley, 1974), 11; emphasis added.

9. Ibid., 11.

10. More is known about the behavior of cercopithicines than about any other group of primates, with the exception perhaps of the chimpanzee. Systematists have debated the taxonomy of the subfamily Cerpithecinae throughout the twentieth century with little consensus. This is especially true of the *Papio* species, or rather subspecies. Many primatologists group all the savanna baboons including *cynocephalus* (yellow baboons), *ursinus* (chacma) and *anubis* (olive) into the species *Papio cynocephalus*. Five species of mangabeys (*Cercocebus*) are closely related to the baboons and live in Africa. In Asia, the macaques consist of sixteen to nineteen species belonging to one diverse genus. See Don J. Melnick and Mary C. Pearl, "Cercopithecines in Multimale Groups: Genetic Diversity and Population Structure," in *Primate Societies: Studies in Adaptation and Variability*, ed. Barbara B. Smuts et al. (Chicago: University of Chicago Press, 1987), 121–48.

11. Ibid.

12. Ibid.

13. Ibid.

14. Robert Jurmain et al., *Essentials of Physical Anthropology* (Belmont, CA: Wadsworth, 2004), 170.

15. Ibid., 90–93.

16. Haraway's magnum opus *Primate Visions* constitutes the most comprehensive piece of scholarship extant on the history of primatology. In it, she limns the entire history of the discipline within the context of shifting historical, cultural, and scientific cosmologies.

17. During the first phase of postwar primatology, a number of long-term studies laid the foundation of the discipline. Jeanne and Stuart Altman studied baboon ecology and behavior at Amboseli National Park, Kenya; Stuart Altman studied a related cercopithecine, the rhesus macaque, in a captive colony on Cayo Santiago Island in the Caribbean. The Smithsonian project observing howler monkey behavior on another Caribbean Island (Barro Colorado) dates from this period. Research accelerated in the 1960s: Jane Goodall began her observations of chimpanzees in Tanzania in 1965; Thomas Struhsaker and others studied several species of monkeys in the Kibale Forest of Uganda; Sherwood Washburn and his student Irven DeVore studied savannah baboons in the Serengeti National Park, Kenya; Diane Fossey initiated observations of mountain gorillas in 1967 in Rwanda's Parc des Volcans; and Phyllis Dolhinow researched langur monkeys at several sites in India. Most of these studies were descriptive natural histories with few explicit links made to human evolution.

18. Sperling, "Baboons with Briefcases," 206.

19. Ibid. In the same article I have discussed the failure of reductionist-adaptionist models to explain important aspects of primate development and diversity. Both structural functionalism and sociobiology commit the fallacy of affirming the consequent, ignoring the many contingent and epigenetic factors present in the social and physical environments in which primates live. In recent decades, functionalist reductionism in primatology has seemed isolated from important discourses about evolutionary epistemology in other disciplines; primatologists addressing this problem have sometimes found themselves tarred with the brush of "anti-Darwinism" and "antievolutionism." Stephen Jay Gould has written of the frustrations involved in critiquing narrow adaptationism: "A former student of mine recently completed a study proving that color patterns of certain clam shells do not have the adaptive significance usually claimed. A leading journal rejected her paper with the comment: 'Why would you want to publish such nonresults?'" (Steven J. Gould, "Cardboard Darwinism," *New York Review of Books* 33 [1986]: 47–54.)

20. Linda Marie Fedigan, *Primate Paradigms: Sex Roles and Social Bonds* (Montreal: Eden, 1992).

21. Sherwood L. Washburn and Irven DeVore, "Social Behavior of Baboons and Early Man," in *Social Life of Early Man*, ed. Washburn (New York: Viking Fund Publications in Anthropology 31, 1961), 91–105.

22. Nancy Tanner and Adrienne Zihlman, "Women in Evolution, Part 1: Innovation and Selection in Human Origins," *Signs* 1 (1976): 585–608.

23. Thelma Rowell, *The Social Behaviour of Monkeys* (Middlesex, UK: Penguin, 1982).

24. Haraway, *Primate Visions*; Sandra Harding, *The Science Question in Feminism* (Ithaca: Cornell University Press, 1986).

25. Carol McGuiness and Karl Pribram, quoted in David Goldman, "Special Abilities of the Sexes: Do They Begin in the Brain?" *Psychology Today* 12 (1978): 56.

26. Edward O. Wilson, *Sociobiology: The New Synthesis* (Cambridge, Mass.: Belknap Press, 1975).

27. William D. Hamilton, "The Genetical Evolution of Social Behavior, I and II," *Journal of Theoretical Biology* 7 (1964): 1–52; Robert Trivers, "Parental Investment and Sexual Selection," in *Sexual Selection and the Descent of Man*, ed. Bernard Campbell (Chicago: Aldine, 1972), 136–79.

28. Sociobiology Study Group of Science for the People, "Sociobiology: Another Biological Determinism," *Bioscience* 26 (1976): 182–86.

29. Sarah B. Hrdy, *Langurs of Abu* (Cambridge: Harvard University Press, 1977).

30. Sperling, "Baboons with Briefcases," 216.

31. Jeanne Altmann, *Baboon Mothers and Infants* (Cambridge: Harvard University Press, 1980); Shirley Strum, *Almost Human: A Journey into the World of Baboons* (New York: Random House, 1987).

32. Sherwood L. Washburn and Irven DeVore, "Social Behavior of Baboons and Early Man," in *Social Life of Early Man*, ed. Washburn (New York: Viking Fund Publications in Anthropology 31, 1961), 91–105.

33. Shirley C. Strum, *Almost Human: A Journey into the World of Baboons* (New York: Random House, 1987).

34. Irven DeVore, quoted in Duncan M. Anderson, "The Delicate Sex: How Females Threaten, Starve and Abuse One Another," *Science* 86 (1986): 42–48.

35. Linda Marie Fedigan and Shirley C. Strum, "Changing Views of Primate Societies: A Situated North American Perspective," in *Primate Encounters: Models of Science, Gender and Society*, ed. Strum and Fedigan (Chicago: University of Chicago Press, 2000), 3–49.

36. Ibid., 37.

37. See Adele E. Clarke and Joan H. Fujimura, eds., *The Right Tools for the Job: At Work in Twentieth-Century Life Sciences* (Princeton: Princeton University Press, 1992).

38. See Clifford Geertz, this volume.

39. By the late 1970s, the divergence between postmodernism in cultural anthropology and Darwinian fundamentalism in physical anthropology created an increasingly fragmented pedagogy—the more eminent the department, the less likely were graduate students expected to study anthropology's four fields. For physical anthropologists, the status and grant money associated with biology (rather than social science) also encouraged this fragmentation. By the 1990s, the physical anthropologists at Berkeley, for instance, had left anthropology entirely to join the newly organized Department of Integrative Biology (as has increasingly occurred in other major departments). Thus many primatology students today may have little formal knowledge of human culture diversity and complexity. Perhaps this creates a greater sense of confidence in the comparability of monkey, ape, and human interactions than would otherwise be the case.

From Scaling to Simulation: Changing Meanings and Ambitions of Models in Geology

NAOMI ORESKES

Some time in the mid-twentieth century, the word *model* in geology changed its meaning. For more than a century, the word had referred to a physical model: a small-scale representation of an earth feature or process. Mimetic models were typically built as experiments to demonstrate the plausibility of a proposed causal agent. Methodological concerns surrounding these models focused on the problems of scaling (asking whether it was possible to capture the earth's capacities on human scales) and the problem of the method of hypothesis (asserting that demonstrating the potential of a causal agent did not prove its actual existence or effect).

Beginning in the 1960s, and accelerating in the 1970s and 1980s, the word *model* took on a different meaning: a computer simulation. For earth scientists, this is the dominant meaning it holds today. With this semantic change came a shift in the aspirations and ambitions of earth scientists. Computer models were built not so much to demonstrate causal efficacy, but to attempt to predict the future behavior of geological systems. Methodological concerns shifted to the question of model validation: how to confirm or deny the predictions of models that are nonunique and whose predictive output is temporally or physically inaccessible.

Although both types of modeling—physical and numerical—have shared the ambition to access the inaccessible, the goal of predicting the future proved a radical shift in the ambitions of earth scientists, who until this time had generally eschewed prediction. What caused such a radical change in epistemic ambitions? This essay suggests that the demand for prediction was driven to a significant extent by Cold War sponsors of earth scientific research and thus provides a concrete example not only of how the epistemic values of scientific communities change over time but also of how those changes can come in response to the aspirations of scientific patrons.

THE NINETEENTH CENTURY: PHYSICAL
MODELS AS GEOLOGICAL EXPERIMENTS

Many people think of experimentation as emblematic of scientific work, but throughout most of the history of geology, fieldwork has provided the dominant mode of investigation. Following the inspiration of Abraham Gotlob Werner, geologists at the end of the eighteenth century fanned out across Europe and North America to determine the sequence of rock units that recounted earth history, to discover the forms of life recorded in the fossil contents, and to unravel the structural effects that had dislocated them. By the time Charles Lyell published his celebrated *Principles of Geology* in 1830, he was advocating what was already common practice throughout much of continental Europe and North America.[1]

Experimentation played only a modest role in the early history of geology. Because one could not experiment on the earth directly, it took substantial imagination to conceptualize what an experiment in geology might even consist of. Moreover, most geologists lacked access to the materials and equipment that might permit reproduction of earth conditions.[2] For this reason, some of the earliest experiments occurred in chemical rather than physical geology. In the early nineteenth century, a number of geologists and chemists looked to experiments to evaluate the two dominant schools of thought about petrogenesis. Wernerians believed that rocks were precipitated from aqueous fluids; Huttonians argued for solidification from melts and compression at great depths. These general claims were tested via specific experiments: Could limestone be precipitated from water? Could granite be generated by first melting and then solidifying quartz and feldspar? Most early experiments of this kind proved inconclusive, so much so that the historian Sally Newcomb has argued that "rather than providing strong reasons for theory choice, laboratory results might instead bring about a condition of no theory choice."[3]

The most famous examples came from the hand of Sir James Hall (1761–1832), often called the "founder of experimental geology."[4] Hall designed a series of experiments to test (and hopefully confirm) the Huttonian view, but he encountered serious difficulties: quartz was almost impossible to fuse; sand was *not* readily compressed into sandstone. Indeed, his attempts to affirm sandstone formation by heat and compression failed. As Newcomb explains, they "ended by his needing to add salt crystallized from water as a cementing agent and nearly put Sir James in the Wernerian Camp."[5]

Hall defended the value of his experiments, but others hotly contested their meaning. Experiments were considered to entail the method of hypothesis, with all the logical difficulties that implied.[6] One could demonstrate that a proposed cause *could* produce an observed effect, but this did not prove that it had done so in nature. It merely demonstrated its *capacity*. When experiments failed— and Sir James's experiences were scarcely unique—the question was raised as to whether the same effect, applied more vigorously or for a longer period, might have proved efficacious.

Hall tried, therefore, to argue that the real purpose of experiments was to convert a proposed cause into a *vera causa*, a cause in harmony with other causes already known. As he explained in 1815, "We are never more disposed to give credit to a philosophical system, than when we meet with a case of its successful application."[7] But others were doubtful, including James Hutton himself, who realized that with friends like Hall, he might not need enemies. In an attempt to explain away Hall's failed experiments, Hutton effectively disparaged experimentation in general. Hall summarized Hutton's view that "the heat to which the mineral kingdom has been exposed was of such intensity as to lie beyond the reach of our imitation, and that the operations of nature were performed on so great a scale, compared to that of our experiments, that no inference could properly be drawn from the one to the other."[8]

This objection was applied a fortiori to the other principal area of experimentation in nineteenth-century geology: the construction of scale models of mountains and crustal deformation. The origin of mountains proved one of the central questions of nineteenth-century geology, and there was no shortage of hypotheses in response. On what grounds could one differentiate between various theoretical suggestions? Sir James pioneered work in this area with his famous compression box: he placed layers of clay or cloth, representing strata, in a box with movable sides and squeezed them in an attempt to reproduce the folded structures observed in mountain ranges.[9]

Hall's compression box became the prototype for physical models of mountains over the next hundred years. While the question of the origins of mountains waned slightly in the mid-nineteenth century, when geologists focused on stratigraphy and earth history, it waxed again later in the century, when various investigators attempted to reproduce fold structures in scale model experiments. Mimetic models played the role of experiments for geologists trying to understand what causal forces had created mountains.

Scale Models of Mountain Building in the Nineteenth Century

Most orogenic theories of the late nineteenth century were variations on the theme of earth contraction, interpreted as a corollary of Pierre Simon, Marquis de La Place's nebular hypothesis. If the earth shrunk, it must have deformed—

FIGURE I Alphonse Favre's experimental apparatus testing whether Earth contraction could produce folded mountains. A band of clay 25 mm with no supports along the side, placed upon stretched rubber, produces slight wrinkles, ~ 3–4 cm high, when the rubber is released. "The Formation of Mountains (A report on the work of Alphonse Favre)," *Nature* 19 (1878): 104.

but how, exactly? Was horizontal compression a side effect of contraction? In 1878, the Swiss geologist Alphonse Favre reported a series of experiments in which clay layers were placed on stretched rubber, which was then released, and the resulting structures observed. The clay did indeed fold, and in a series of photographs Favre presented the results (figure 1). His conclusion? Horizontal compression *could* be a side effect of contraction, in which case the flexures produced would move largely in an upward direction.

For Favre, the important point was that his model was successfully mimetic, for the structures produced in the model "appear similar to those of hills and mountains which may be observed in various countries."[10] Then he varied the parameters of the model to determine which conditions most closely mimicked nature, and to reproduce some of the variety seen (figure 2, a and b). Favre found that thicker bands of clay produced more conspicuous folds. Varying the thickness across the layer produced asymmetrical folds, and placing cylinders of wood under the clay layer localized the deformation. From the success of his scale model in reproducing a variety of features actually seen in the field and their interrelations—what Favre called "accidents of the ground"[11]—he concluded the affirmation of his hypothesis: "Powerful lateral thrusts of the external and solid parts of the globe appear to result from a diminution which the radius of the interior pasty or fluid nucleus has undergone during millions of ages. It may have been sufficiently great to cause the solid crust (which must have always been supported on the interior nucleus, whose volume continually diminishes) to assume the forms which we know, with a slowness equal to that of the contraction of the radius."[12] In other words, the scale model supported the *plausibility* of the hypothesis.

FIGURE 2 The effect of varying parameters: (a) varying the thickness across the layer produces asymmetrical folds; (b) placing cylinders of wood under the clay layer localizes the deformation. "The Formation of Mountains (A report on the work of Alphonse Favre)," *Nature* 19 (1878): 104.

The following year, in 1879, the French geologist and the director of the École Nationale des Mines, Auguste Daubrée (1814–96) published his *Études synthétiques de géologie expérimentale*, in which he presented the results of thirty years of research (figure 3). Daubrée was an ambitious advocate for the experimental method, which he felt was needed to transform geology from a passive science based on "observation of natural facts, accompanied by reasoning and induction," to a positive one providing "certainty or true demonstrations."[13] Experiments could do this because they consisted of "active observation," through which scientists became not just observers after the fact but "witnesses in each instant" (9). This was perhaps a response to William Whewell's famous complaint that we "know causes only by their effects";[14] for Daubrée, experiments allowed us to witness causes through "positive experience," permitting geology—like physics and chemistry—to arrive at "incontestable principles" (10).

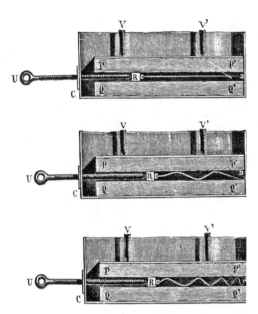

FIGURE 3 Auguste Daubrée's compression box and two experiments:
(a) compression of metal plates produces simple symmetrical folds;
(b) varying orientations produce asymmetrical folds. Auguste Daubrée,
Études synthétiques de géologie expérimentale (Paris: Dunod, 1879), 4.

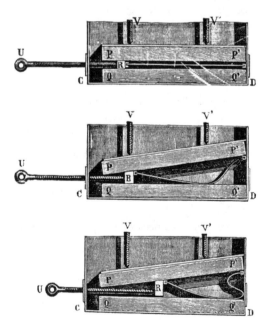

Despite his faith that scale models provided the geologist with positive ex-periences, Daubrée acknowledged that these experiences were problematic when it came to mechanics, where the processes could not be faithfully repro-duced for reasons of scale. "The equipment and forces that we can set to work are always circumscribed, and they can only imitate geological phenomena at the scale ... of our own actions," he admitted (5): "The great fractures and folds that reveal themselves throughout the earth's crust, in mountain chains and elsewhere, are extremely difficult to access in experience, above all because of their great dimensions" (288). Scale models of mountains were therefore not rigorous demonstrations, and certainly not replications, but they did prove useful in illuminating possibilities: "The resemblance in these effects may make us suppose a certain analogy in the causes" (346).

Daubrée's work was closely followed by the British geologist T. Mellard Reade. In *The Origin of Mountain Ranges, Considered Experimentally, Structur-ally, and Dynamically* (1886), Reade focused on vertical uplift rather than hori-zontal compression. Following the American geologist James Dwight Dana, Reade proposed that mountains resulted from surface uplift, caused by thermal expansion in the underlying sedimentary pile. While the effects of any given episode of thermal expansion might be small, large effects could be produced by "small recurrent causes."[15]

To demonstrate this, Reade performed a series of experiments on natural materials to calculate the expected volume expansion of buried sediments ex-posed to natural geothermal gradients, using a value of 1° Fahrenheit per sixty feet based on measurements from mines and boreholes. His began with a strip of lead, one inch wide and fastened at each end, and demonstrated that the strip did indeed bow upward on heating (figure 4). More important, it retained its deformation on cooling. "The lead, although free to contract, does not regain its original form," he wrote.[16] Transient thermal effects could produce lasting deformation.

Following his experiments with lead sheets, Reade worked with three kinds of sandstone, a block of oolitic limestone, and two varieties of marble, each of which produced similar effects. He also measured the lateral elongation of materials under heating and showed that elongation persisted after cooling to room temperature, after seven days, and after fourteen months. While the ob-servable effects were modest, they made the point "that rock is not the immo-bile material we are apt to think it is."[17]

Reade was also interested in the role of lateral compression in generating folds, but he felt that previous experiments were unrealistic because they in-volved only localized compression at the edges of the experimental apparatus,

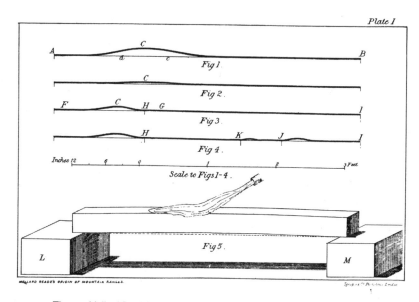

whereas on earth compression would be experienced globally, not locally. This was the subject of Reade's second book, published in 1903, *The Evolution of Earth Structure*. In an innovative series of experiments, he tested the response of model strata to general compression from a zinc band two inches wide (figure 5).

Reade's circular compressor mounted a direct challenge to his fellow experimenters Bailey Willis and Henry Cadell. In 1890, Cadell had published a widely noted article, "Experimental Researches in Mountain Building," in the *Transactions of the Royal Society of Edinburgh*, proposing a mechanism for the formation of imbricate faults. Cadell had worked with the British Geological Survey in the northwestern Highlands of Scotland under the direction of Benjamin Peach and John Horne, where field mapping had revealed great overthrusts displacing large stratigraphic thicknesses over enormous horizontal distances.[18] How in the world did this happen? With the approval of the director of the survey, Cadell embarked on a series of experiments with scale models to "throw light on the [field] work by seeking to imitate in the laboratory the processes we believe to have been in operation in our wild Northwest Highlands" (figure 6).[19]

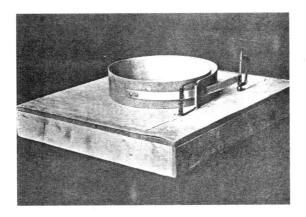

FIGURE 5 Thomas Mellard Reade's circular compressor, designed to simulate the effects of earth contraction; (a) the device; (b) the results of compression of clay layers. Thomas Mellard Reade, *The Origin of Mountain Ranges Considered Experimentally, Structurally, Dynamically, and in Relation to their Geological History* (London: Taylor and Francis, 1886), 27.

FIGURE 6 Henry Cadell with his compression box. Henry Cadell, "Experimental Researches in Mountain Building," *Transactions of the Royal Society of Edinburgh* 35 (1888): 337.

The most distinctive feature of the Northwest Highlands was the evidence that the rocks had "snapped" rather than folded, so Cadell's experiments were designed to replicate brittle behavior. Following Favre and Daubrée, Cadell constructed a compression box, but rather than use clay or wax for the layers, he used plaster of paris interleaved with colored sand. Cadell was pleased with the results, writing, "The researches were attended with marked success [because] the structures obtained showed a striking similarity to those observed in the field. . . . The accompanying figures tell their own tale, and require but little description."[20]

The experimental results indicated that the field relations could be easily misconstrued: strata were repeated not because of a cyclic depositional environment as most geologists would presume, but because faults had stacked one end of a sequence on top of the other end (figure 7).[21] The experiment both confirmed the plausibility of the field relations and served as a guide to them. Cadell concluded: "It is obvious that beds repeated in this way in nature, without inversion or folding, might come to have an appearance of an enormous thickness, and thus greatly mislead the field geologist. . . . This structure is now known to be of common occurrence in the Northwest Highlands, and these experiments show clearly on what mechanical principles many of the extraor-

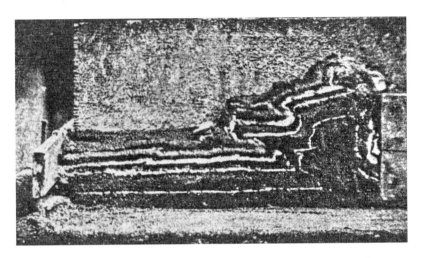

FIGURE 7 Close-up of Cadell's experiment, illustrating the progressive development of implicate faults. Note how the layers are displaced across apparently planar, dipping surfaces. Henry Cadell, "Experimental Researches in Mountain Building," *Transactions of the Royal Society of Edinburgh* 35 (1888): 337.

dinary and remarkably deceptive relationships of the rocks of that region may be explained."[22]

Like Cadell, Bailey Willis (1857–1949) sought to account for features observed in the field, in this case the American Appalachians, which he had mapped as a geologist for the U.S. Geological Survey. In 1894, Willis published the results of his experiments with a compression box in order to "permit a discussion of the laws developed by experimental study and their application to natural phenomena."[23] Unlike Cadell, Willis used soft materials in his experiments, arguing that geological observation demonstrated that, while the saying was "hard as a rock, in resistance to the forces of the earth's mass, this same rock may be relatively as soft as wax" (219).

Willis directly confronted the criticism that the ability to imitate natural structures did not prove an understanding of "either the imitation, or through it the original. For this reason, some have cast experiments aside as useless, and others have been content [merely] to describe their unexplained results" (230–31). Willis believed he could do better because the laws of nature would apply equally well on the scale of the earth as on the scale of a compression box. "Mechanical laws do not vary with the magnitude of the active forces nor with that of the passive resistance," he wrote; "of a series of strata 100s of feet thick and of a pile only inches thick, the bending, breaking, or shearing will obey the

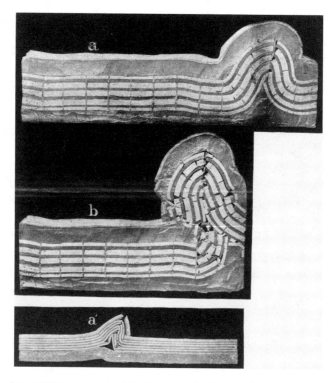

FIGURE 8 Bailey Willis's compression box, illustrating the formation of ductile folds. Bailey Willis, "Mechanics of Appalachian Structure," *Annual Report of the United States Geological Survey, 1891–1892* 13 (1893): 219.

same laws, if all the factors of pressure and resistance are proportionate in each case" (219). But how did one know if those factors were proportionate? Willis reasoned that it was appropriate to use soft materials that would yield easily in the short run just as rocks yield readily in the long run.[24] In the earth's interior, he wrote, brittle fracture was "impossible for want of space, [and] rocks must change form by 'plastic flow'"—an idea he labeled "latent plasticity."[25] The key to an adequate scale model was to create layers pliable enough to behave in a plastic manner yet strong enough to transmit stress, so he blended beeswax with varying proportions of plaster of paris and turpentine to achieve a consistency he called "plasticity without weakness."[26]

Willis's apparatus consisted of an oak box (three feet long, six inches deep, and one inch wide) with removable sides and a screw piston (figure 8). He varied the parameters of the layers to try to determine the influence of strata thickness, of loading, and of variations in plasticity. The results proved consistent: an anticline always formed on the side nearest the piston. In the presence of inaugural irregularities, folds initiated on the side of the irregularity. When irregularities occurred in more than one layer, those in the firmest layer controlled the fold location. He called this the "competent layer," referred to the pattern as the phenomenon of "competent development," and summarized his results in the "law of anticlinal development": "In strata under load an anticline arises along a line of initial dip, when a thrust sufficiently powerful to raise the load is transmitted by a competent stratum."[27] For Willis, experiments were guides to laws of nature otherwise inaccessible.

SCALE MODELS IN THE TWENTIETH CENTURY

While field geologists at the end of the nineteenth century considered it indisputable that rocks under pressure behaved in a plastic manner, seismologists had become equally convinced on the basis of shear wave transmission that the earth's interior was rigid. This paradox set the terms for heated arguments in the early twentieth century over the strength of the earth and the possibility of continental drift. It also led to further experiments with scale models, of which the most well known were those of the Dutch geologist Philip H. Kuenen and an American geophysicist named David Griggs.[28]

Philip Kuenen was one of the most versatile geologists of the twentieth century. Having begun his career in the Dutch *Snellius* expedition to the East Indies in 1929 and the 1930s, he maintained a lifelong interest in marine geology, submarine topography, vulcanism, and sedimentation, and he built numerous scale models of these features and processes in an attempt to understand them. Most famously, Kuenen is credited with the experimental proof of the existence of turbidity currents (submarine mudflows) and their role in the formation of submarine canyons.

In the 1920s, the mapping of submarine canyons with the newly invented echo sounder led to a debate about their origins.[29] At the Scripps Institution of Oceanography, in La Jolla, California, the marine geologist Francis Shepard had proposed that they were drowned river valleys, inundated at the end of the last ice age. Using bathymetric data, Shepard showed the canyons had a characteristic morphology—elongated perpendicular to the shoreline—and that they were comparable in depth and pitch to the largest subareal river canyons (figure 9). Furthermore, many were located near the mouths of terrestrial rivers, so it was plausible to think of them as submarine extensions of them. At Harvard, however, the well-known geology professor Reginald Daly argued that the canyons

FIGURE 9 Philip Kuenen's scale model of submarine canyons on Georges Bank. Philip H. Kuenen, "Experiments in Connection with Daly's Hypothesis on the Formation of Submarine Canyons," *Leidsche Geologische Mededeelingen* 8 (1937): 327–51, based on hydrographic charts by the Scripps Institution of Oceanography marine geologist Francis Shepard.

had been carved out by submarine mudflows, mainly because the amount of eustatic sea level rise that would have been required to submerge the canyons was ten times that normally assumed by glacial geologists.[30]

Daly had written that "on account of the difficulty of scale, convincing experiments with laboratory models are not to be readily performed."[31] Kuenen disagreed, arguing that experiments could at least help constrain possibilities: Will a suspension flow down a slope under water? Will it maintain its internal coherence over distance? Would it have erosional power? Would it deposit sediments resembling those found in nature? Kuenen did not claim that his experiments provided decisive proof, but he did note that "several geologists who witnessed the experiments admitted being more favorably disposed towards the theory than at first."[32]

Kuenen constructed a flume tank with a base of sand to simulate the sediments on the seafloor, covered with a layer of gypsum for resistance, and set at a moderate incline (figure 10). Then he created a suspension of mud and water, which he poured behind a wooden dam. When the wood was lifted, the slurry flowed downhill. All of Kuenen's questions were immediately answered in the affirmative: the mudflow retained its integrity; it had erosional power, carving out a long narrow canyon whose morphology was comparable to that of the canyons; and the sediments deposited at the end closely resembled greywacke deposits found interbedded with deep sea deposits in nature. Most spectacularly, these so-called turbidites reproduced the characteristic fine-scale struc-

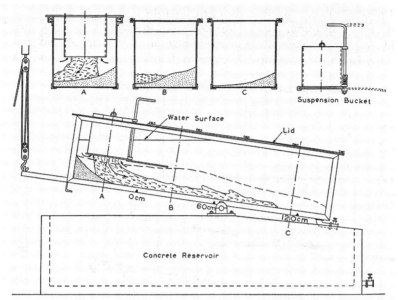

Fɪɢ. ɪ.—Tank with suspension bucket as used for studying small-scale turbidity currents of high density

FIGURE 10 Schematic of Philip Kuenen's flume tank, designed to test the possibility of the existence of turbidity currents in nature, and their role in the formation of greywackes. Philip H. Kuenen, "Experiments in Connection with Daly's Hypothesis on the Formation of Submarine Canyons," *Leidsche Geologische Mededeelingen* 8 (1937): 327–51.

tures of greywackes, including ripple marks, sole marks, and finely graded beds (figure 11).

Kuenen continued work on turbidity currents for two decades; he also contributed to discussions of crustal structure through scale models of tectogenes. In the early 1930s, the American geologist Harry Hess had developed the idea of a large down-warping of the crust to explain the large negative gravity anomalies discovered by the Dutch geodesist Felix A. Vening Meinesz. Hess had worked with Vening Meinesz on gravity measurements in the Caribbean, and suggested that the negative anomalies might be explained by enormous downward protrusions, in which low density crust was folded into the higher density mantle below. Using a term introduced by the German geologist Erich Haarman, Hess called these down-warpings "tectogenes."[33]

Kuenen constructed an experimental apparatus to test the idea: a glass aquarium filled with layers of paraffin, Vaseline, mineral oil, and warm water, and equipped with beams at either end to generate horizontal compression. De-

FIGURE II Kuenen's flume tank experiment. In the upper photographs, a simulated turbidity current has been produced. In the lower photograph, ripple marks appear in the sediments deposited from the simulated turbidity current. Philip H. Kuenen, "Experiments in Connection with Daly's Hypothesis on the Formation of Submarine Canyons," *Leidsche Geologische Mededeelingen* 8 (1937): 327–51.

pending on the strength of the layers and the amount of compression, he could produce either crustal down-warpings, as proposed by Hess, or underthrusts, as proposed by European geologists such as Albert Heim and Otto Ampferer (figure 12). Kuenen did not claim his models proved that tectogenes existed, but they did demonstrate that they *might*.

Advancing the Science: Physically Similar Scale Models

Kuenen described his models as heuristic devices, useful for "checking . . . theoretical deductions and guiding further research."[34] David Griggs, a professor of geology at the University of California, Los Angeles, agreed. Griggs was also interested in crustal deformation and continental drift. By the mid-1930s, the theory of drift had received wide attention in the United States: many believed that the gravity anomalies measured by Vening Meinesz and Hess resulted from crustal down-warpings caused by lateral compression, and Arthur

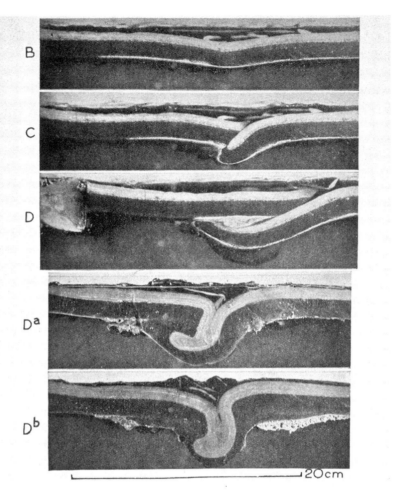

FIGURE 12 Kuenen's scale model of a tectogene. In B-C-D, as compression increases, one half of the "crust" slides beneath the other. In D^a and D^b, the experiment is sliced open to reveal a deeper "root" inside the "crust." Philip H. Kuenen, "The Negative Isostatic Anomalies in the East Indies (with Experiments)," *Overdrukuit Leidsche Geologische Mededeelingen* 8 (1926): 170–71.

Holmes had proposed that convection currents in the mantle were the driving force. Griggs set out to test this complex of ideas by an experimental apparatus that mimicked Holmes's idea.

The apparatus consisted of two rotating drums in a "sea" of oil, covered by a floating layer of paraffin (figure 13a). When the drums were moved, the surface layer was stretched until it migrated across the surface. When one drum was rotated and the other kept stationary, it produced a thickening and down-warping of the crust consistent with Hess's proposal and overall crustal mobility (figure 13b). Griggs called his setup "a dynamically similar scale model" because he had attempted to scale the viscosity of the materials in accordance with the principles articulated by his American colleague M. King Hubbert, one of the first earth scientists to develop a quantitative treatment of the scaling problem.[35]

Hubbert, the associate director of exploration and production for Shell Oil, was the most famous oil company geologist in the United States. In 1937, he had published a widely read article, "Theory of Scale Models as Applied to the Study of Geologic Structure," based on his PhD work at Columbia, which addressed the problem of scaling in physical models.[36] This was followed by a second article in 1945, "Strength of the Earth," in which he addressed one of the "most perplexing problems in geologic science," namely, "how an earth whose exterior is composed of hard rocks could have undergone repeated deformations *as if* composed of very weak and plastic materials." Seismic evidence and tidal phenomena suggested a rigid earth; isostasy and field evidence suggested a weak earth. Hubbert argued that the difference was "more apparent than real" and the principles of scaling showed why.[37]

Hubbert's was the first fully quantitative treatment of the question how to choose the physical properties of materials in a model to account for the much smaller scale and time frame as compared with nature. Like Daubrée, Hubbert was optimistic that a properly scaled model could enable scientists to understand events and processes otherwise beyond not only human experience but even our imaginative capacity: "By means of the principles of physical similarity," he wrote, "it is possible to translate geological phenomena whose length and time scales are outside the domain of our direct sensory perceptions into physically similar systems within that domain" (1630).

Hubbert illustrated the point by a thought experiment (figure 14): Imagine you were to quarry a single block the size of the state of Texas. Its thickness is one-fifth its width, and it is perfectly monolithic and flawless. Would the rock be strong enough to quarry, without breaking apart? While a Texan might attempt such a feat, an easier approach would be to build a scale model. But, what

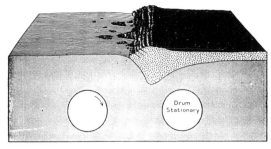

Fig. 15. Stereogram of Large Model with only One Drum Rotating, Showing Development of Peripheral Tectogene.

THE MOUNTAIN BUILDING CYCLE

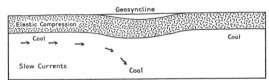

1. First stage in convection cycle — Period of slowly accelerating currents.

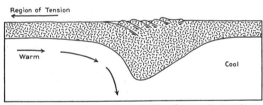

2. Period of fastest currents — Folding of geosynclinal region and formation of the mountain root.

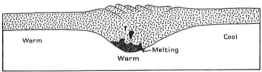

3. End of convection current cycle — Period of emergence. Buoyant rise of thickened crust aided by melting of mountain root.

Fig. 16. Hypothetical Correlation between Phases of the Convection-Current Cycle and Phases of the Mountain-Building Cycle. Structural Relations Drawn from the Model.

FIGURE 13 David Griggs's model of subcrustal convection as a driving force of continental drift: (a) an illustration of the scale model setup; (b) the tectonic interpretation, with "structural relations drawn from the models." David T. Griggs, "A Theory of Mountain Building," *American Journal of Science* 237 (1939): 643, 645. Reprinted with permission of the *American Journal of Science*.

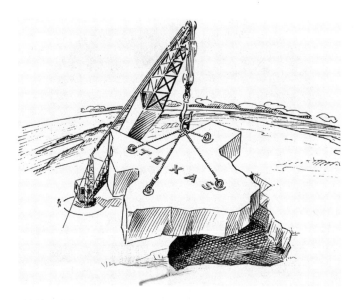

FIGURE 14 M. King Hubbert's thought experiment: quarrying Texas to determine the strength of crustal materials. M. King Hubbert, "Strength of the Earth," *Bulletin of the American Association of Petroleum Geologists* 29 (1945): 1634. Copyright AAPG 1945, reprinted with permission of the AAPG whose permission is required for further use.

materials, of what strength, would one need to use in the model for it to be physically similar to Texas? A little consideration reveals that we already know the answer. A map is geometrically similar to the landscape it represents if the ratios of distances on the map are equal to the ratios of the same distances on the landscape. That ratio is the scale of the map. In a physical model, the ratios of the model properties—length, width, density, viscosity—must likewise retain the same ratios as in nature. These are the scaling factors of the model.

For Texas, if all factors were constant except the length of the block, then the strength of the material must be reduced proportionately to the length reduction. If the scale model were 60 cm long, then the strength reduction would be 5×10^{-7} (60 $\times 10^{0}$ cm / 12 $\times 10^{7}$ cm = 5×10^{-7}). That is, the strength of the materials in the scale model must be seven orders of magnitude less than the strength of crustal rocks. For bench-scale models of crustal deformation, the materials would need to have the consistency of "very soft mud or pancake batter" (1651). In other words, it would be a solid that behaved like a liquid, something outside of our ordinary experience.

With the exception of Griggs, who had followed Hubbert's guidelines, previous investigators had all used inappropriate materials. For example, Willis's model strata had fractured extensively and broken into discrete slabs and open cavities, but in nature open cavities in folded rocks are not observed. The mistake was now obvious: Willis's materials were too rigid. In contrast, Griggs's results "were in remarkable accord with the major tectonic features of many existing mountain ranges" (1653). Hubbert's example was based on physical similarity in space, but given that elastic properties also depend on strain rate, the same argument would apply in time. A scale model that operated orders of magnitude faster than terrestrial processes would have to use materials that were orders of magnitude weaker than terrestrial rocks.

Hubbert's conclusion presaged an important development in the earth sciences. If one could calculate the required properties of materials in a scale model, then there was actually no need to build the model itself. One could simply calculate the property of interest. In the case of arguments over the strength of the earth, Hubbert's point—that the earth behaved as if it were pancake batter over appropriate ranges of time and space—answered the question at hand without recourse to the physical model it was intended to guide. Hubbert's article was published in 1945; soon computers would be available to do the calculations for him.

FROM SCALE MODELS TO SIMULATIONS

The scale models that earth scientists built over the course of nearly three centuries involved a variety of configurations mimicking earth features, but they all shared a common feature: each constituted an attempt to access the inaccessible. As Daubrée argued, we cannot witness the processes that cause the effects we observed, but an experiment on a scale model could provide us with a "positive experience." No wonder so many scale models addressed crustal deformation: this perhaps constituted the most important causal question in the earth sciences that was incapable of direct observation. But the same held true for models of volcanoes and sedimentary processes. The goal of the investigators who built physical models was to access the inaccessible. Modeling in geology constituted an attempt to know causes not just by their effects but by reproducing them in real time.

The use of computer models in the earth sciences in the past few decades has shared the goal of accessing the inaccessible. In principle, a computer simulation can be used in precisely the same manner as a mimetic physical model: to demonstrate circumstances capable of producing known effects. Sometimes simulations have been employed in this way. In seismology, simulations have been used to determine what mix of internal structure and properties can account for the known transmission of seismic waves; in geochemistry, to analyze the processes responsible for mass transport and mineral deposition; in hydrology, to analyze large-scale groundwater flow systems. But in investigating inaccessible areas and aspects of the earth, numerical simulation models have been associated even more strongly with another ambition: the desire to predict the future.

It might seem obvious that geologists would want to make predictions: most scientists do.[38] Furthermore, the logical-empirical model of science widely promulgated in the mid-twentieth century, and widely accepted by scientists even today, makes prediction a sine qua non of theory testing. We might even be tempted to "predict" that earth scientists would want to make predictions, and seek the means to do so. Historical evidence suggests a different reading. While earth scientists in the early twentieth century wanted to make their science more quantitative and more experimental, the impetus for prediction came largely from postwar patrons of the earth sciences. That is to say, the demand for prediction came largely from outside the scientific community.

Until the time Hubbert published his "Strength of the Earth," virtually no activity within the geological community existed that could reasonably be called predictive. Throughout the eighteenth and nineteenth centuries, and well into the twentieth century, most geologists were engaged in the enterprises of explaining the past or investigating the present. These tasks were understood as explanatory and retrodictive. The present was important for explaining the processes operative in the past, and the past was the special domain of earth scientists.

As the philosopher David Kitts argued some time ago, there was little meaningful sense in which geologists might even think to make predictions because geology for the most part lacked laws.[39] This was a consequence of geology's character as a historical science dealing with singular events. Geological attempts to formulate laws—such as Willis's "law of anticlinal development" noted above—rarely proved successful.[40]

There are two major exceptions to this generalization within the broader earth sciences: seismology and meteorology. Some seismologists in the early twentieth century expressed a general hope that it might one day be possible to predict and even prevent earthquakes like the one that devastated San Francisco in 1906, but this was a hope, not an expectation. Most seismologists were far too busy with instrument development and basic theoretical questions to imagine that they were anywhere close to achieving that desideratum.

The other exception is meteorology. As both the historians James Fleming and Robert Marc Friedman have documented, scientific weather forecasting has a long history, one supported by its obvious value in navigation and military campaigns. Empirical forecasting received a large boost in the nineteenth century by the development of the telegraph, making it possible to transmit information about weather in adjacent areas and thereby to recognize patterns.[41] In the early twentieth century, the Norwegian physicist Vilhelm Bjerknes pioneered the development of quantitative meteorology with impetus from the fledgling aviation industry.[42] During World War II, weather forecasting was advanced by the U.S. Army Weather Directorate, which sponsored a major predictive effort for the D-day landing and the analysis of coupled hydrosphere-atmosphere interactions for the forecasting of surf conditions during amphibious assaults.[43]

These activities presaged a greatly increased military interest in the earth sciences during the Cold War period, but they nevertheless involved only a portion of the earth sciences community; few geologists participated. However, a sea change occurred in the late 1960s and early 1970s, with the demand for prediction of the long-term behavior of proposed nuclear waste disposal sites.

Charles D. Hollister and the Subseabed
Disposal of Nuclear Waste

The problem of the disposal of nuclear waste from the nation's new nuclear weapons program was born with the Manhattan Project, but it did not gain much attention until 1954, with the passage of the Atomic Energy Act,[44] which made high-level waste the responsibility of the federal government. As the U.S. government dramatically increased the scale of its weapons program, and began to foster the development of civilian nuclear power generation, the anticipated quantity of waste soared. Earth scientists became involved in research projects designed to evaluate potential waste disposal schemes.

This work accelerated in the 1960s and 1970s as the Atomic Energy Commission (AEC) began to evaluate potential sites for geological disposal, both on land and at sea. It was further boosted in 1982 with the passage of the Nuclear Waste Policy Act, which required the U.S. Department of Energy (DOE) to establish a · process for screening proposed repository sites and to establish a schedule for waste acceptance by 1998.[45]

The Atomic Energy Commission (the Department of Energy after 1974) contracted large numbers of earth scientists to evaluate the suitability of proposed nuclear waste disposal sites. Because nuclear waste would remain toxic and radioactive for at least tens of thousands of years, geological sites were felt to offer the best chance of containment. To affirm the suitability of a site was to predict its behavior over the next ten thousand years, and AEC/DOE contracts explicitly required geologists to make such predictions. Often, these predictions came to be framed as the output of models—initially conceptual ones, later numerical simulations. This is a large subject, the details of which are well beyond the scope of this study, but one example will suffice to illustrate the point.

Charles D. Hollister (1936–99) was a marine geologist who obtained his PhD at Columbia's Lamont Geological Laboratory, working under Bruce Heezen. Heezen had a long and highly productive career investigating deep ocean structure and sedimentation, built largely on the financial support from the Office of Naval Research (ONR). Together with Marie Tharp, Heezen produced the first physiographic maps of the ocean floor, made famous by their publication in *National Geographic*.[46] In the late 1960s, Hollister worked with Heezen on a project investigating deep-ocean sedimentation, of interest to the U.S. Navy for the potential impact of deep currents on underwater hydrophone arrays used for acoustic surveillance of Soviet submarines. Hollister was one of several scientists working at Lamont on ONR funding to demonstrate—contrary to prevailing wisdom—that vigorous currents scoured the seafloor along the continental slope, transporting and depositing large amounts of clastic sediments.[47]

The discovery of contour currents—named for the fact that they tended to run parallel to topographic contours—had serious negative implications for proposals to dispose of nuclear waste at sea. If the continental slopes were swept by powerful currents, then any attempt to dispose of nuclear wastes would require materials to be taken further out to sea, perhaps to the abyssal plains. This meant greater expense and scientific uncertainty, for the abyssal plains were even less explored than the continental slopes. However, at just this time, the establishment of plate tectonics indicated that the interior regions of oceanic plates might be tectonically stable for tens to hundreds of millions of years. The AEC now took an interest in the prospect of the seabed disposal of nuclear waste in the stable interiors of oceanic plates.

After completing his PhD, Hollister took a job at the Woods Hole Oceanographic Institution (WHOI), where he continued his work on deep-sea circulation and sedimentation. Beginning in the early 1970s, and continuing throughout the following decade, he received extensive funding for research on the marine disposal of nuclear waste. Hollister directed a study group at WHOI known as the "Deep sea geology study program," and served as the Coordinator for the DOE's U.S. Seabed Disposal Program Site Suitability/Selection programs. He also served as the task force leader of site selection for the DOE's International Seabed Working Group, which examined potential sites outside U.S. jurisdiction. Finally, he was asked to identify and evaluate proposed sites in both the Atlantic and Pacific Oceans.

A research contract dated November 18, 1978, elaborates on the nature of his work and the information to be supplied. Hollister was asked to evaluate a proposed site in the eastern North Atlantic, and to begin investigations that might lead to the identification of potential sites in the south Atlantic and south Pacific. The latter study was to consist of the analysis of historical data to determine the tectonic, volcanic, and sedimentological stability of these regions.[48]

A major part of the project was the development of a paleo-environmental/predictability model of a site, based on information obtained from deep-sea core samples. The model was to encompass the behavior of the western north Atlantic over the past ten thousand to one hundred thousand years based on the mineralogical and paleontological characteristics of the sediment samples. From this, Hollister was asked to predict the behavior of the site over the next 1 to 10 million years, or as the contract put it, "the next 10 to the 6th power years." The contract explained: "Credible prediction of a stable . . . environment is one of the central issues needed for the assessment of a site's suitability as a repository. This predictive effort can only be founded on a complete record of the depositional environment over the past 10,000 to 1,000,000 or more years. The paleo-environmental/predictability model work being done under this task will continue to be the basis for assessing a site's usefulness as a repository."[49] As far as I have been able to determine, Hollister had never claimed that the study of a region's geological history *could* generate reliable predictions about its future, but he did not decline the task set before him. Hollister's model was conceptual in a traditional geological way: it attempted to forecast the likelihood of active tectonism or vulcanism in the region based on the past occurrence of such events as recorded in the sediment pile. However, it involved a tacit quantitative dimension because the model would "address the probability of all natural environmental perturbations that could affect the regions suitability as a possible repository site."[50] While the contract did not explicitly state how such probabilities would be calculated, it implied that they would be extrapolated from the frequency of such perturbations in the past, as indicated by the sedimentological record.

Within a few years, the focus of policy attention had shifted to land-based disposal, and with it the focus of scientific concern to groundwater flow.[51] By the 1980s, the models sought were no longer merely conceptual but became expressed in the form of numerical simulations. Since the mid-1980s, hundreds of American earth scientists have been involved in the development and evaluation of computer simulations to evaluate proposed repositories and natural analogs. With the growth of computer modeling as a methodological tool, its use has spread to other areas of the earth sciences, and programs originally developed in aid of the waste disposal program are now routinely used for other purposes as well.[52]

WHAT EPISTEMIC WORK DO THE MODELS DO?

Just as scientists in the eighteenth and nineteenth centuries had argued about the meanings of scale models, so scientists in the twentieth century argued about the meaning of numerical simulation models. One of the complaints was similar: a numerical simulation model could no more prove that a proposed cause had actually occurred than could a physical model. But with the use of models to predict the future, a new set of epistemic concerns arose: how to confirm or deny the predictions of models that are nonunique and whose predictive output is temporally or physically inaccessible.

Unlike the physical models of previous centuries, which typically contained only a handful of physical components, numerical simulation models may include dozens, scores, or hundreds of equations, scores or hundreds of input parameters, and thousands of lines of code. If the model fails to match observations made of the natural world, how does one decide where the problem lies? Scientists typically refer to this problem as nonuniqueness or equifinality; philosophers know the problem as empirical equivalence and underdetermination. Given a complex model with many parameters, there can be an almost infinite number of arrangements that will permit the model to match observed empirical information.

A second problem, specifically arising from predictive aspirations, is what to do with a model whose predictions are temporally or physically inaccessible. Many earth science models address questions on large temporal scales: climate change over the next century, radioactive waste disposal over the next ten millennia. In such cases, the predictions cannot be tested, because by the time the actual effects are known, the results will be superfluous. In the case of climate change, we will live to see the effects of increased atmospheric carbon dioxide, but by the time we know the full extent of those effects, it may be too late to do anything about it. In the case of radioactive waste disposal, none of us will be around to know what actually happens. Other models address large physical scales, such as the circulation of the world's oceans and atmosphere or groundwater transport across central Australia. In these cases, model predictions may also be inaccessible because we do not have access to the systems on the scale at which the model models them.

In response to these difficulties, modelers have developed a variety of smaller-scale tests that frequently go under the name of "model validation." The idea is that tests of the model on a small scale can demonstrate the legitimacy of the model at a larger scale. But the difficulties with this concept are many. First, systems may have emergent properties not evident on smaller scales. Second, small errors that do not impact the apparent fit of a model to available data may accumulate over time and space in ways that would affect the fit over the long run. Third, models that predict long-term behavior may not anticipate changes in the system over time—such as changes in boundary conditions or forcing functions—that may radically alter system behavior.

These issues have been the subject of extensive discussion among modelers over the past two decades; none of them have been resolved.[53] Rather, as was the case in earlier periods, scientists continue to construct models despite the difficulties they entail, in the belief or hope that they will in some way prove informative. In this sense there is continuity with earlier practice. However, in the sense that numerical models attempt to predict the future, we also find an important discontinuity. The historical point to be made here is this: earth scientists did not take on the task of prediction solely (or even primarily) for epistemological reasons. A significant share of the demand for credible prediction in the earth sciences was generated by the social and political context of nuclear waste disposal and other policy issues. The task of geologists was to build predictive models that could be used to say yea or nay to a proposed repository site. Similarly, the task of much climate modeling, hydrological modeling, geochemical modeling, etc., has been to generate predictions to inform policy decisions.

In Charles Hollister's case, predicting the future behavior of a proposed marine repository site did not add to his knowledge or understanding of it. On the contrary, the predictions he made were so far in the future that neither he nor any of his students would be around to tell whether or not they came true. In this sense, the sorts of predictions earth scientists became involved with differed from the logical predictions imagined by the hypothetico-deductive model of science. In the latter case, predictions play a crucial role in the formulation of scientific knowledge *because* they provide a means to test it. But in the former, predictions serve the goals of patrons, irrespective of whether they contribute to the scientific basis for those predictions. Predictions in geology now serve a role that is primarily social, rather than epistemic.[54]

CONCLUSION: TO ACCESS THE INACCESSIBLE

The predictive models of the late twentieth century and the physical models of earlier times have shared the goal of accessing the inaccessible. Physical models attempted to access causes acting in the past or present; conceptual and simulation models attempt to access the future. In this respect, we might argue that the shift did not prove fundamental, that the various models merely represented different means of achieving the same goal.

Yet I would argue that such a position would be mistaken, for in taking on the future, earth scientists accepted a radical shift in the ambitions of their science. Even today, many geologists deny the appropriateness of making forecasts of future conditions. When the geologist Charlie Rubin was asked a few years ago if his work on the frequency of ground motions along the San Andreas Fault would permit geologists to predict the next earthquake in southern California, he replied: "I'm a geologist, I don't make predictions."[55] While Rubin's comments may reflect the persistent impulses and instincts of many geologists, political forces have moved the discipline in a contrary direction. The shift toward prediction in the earth sciences was not primarily a manifestation of the epistemological aspirations of earth scientists. On the contrary, it went against the grain of two hundred years of geological practice. The rise of prediction was a response to the needs and aspirations of patrons. Prior to the development of numerical simulation techniques, earth scientists rarely made predictions. Indeed, they considered it outside the scope of their discipline. With the advent of computer simulation, predictive modeling in the earth sciences has now become widespread.

NOTES

1. Alexander M. Ospovat, "The Distortion of Werner in Lyell's *Principles of Geology*," *British Journal for the History of Science* 9 (1976): 190–203; Rachel Laudan, *From Mineralogy to Geology: The Foundations of a Science, 1650–1830* (Chicago: University of Chicago Press, 1987); David Oldroyd, *Thinking about the Earth: A History of Ideas in Geology* (London: Athlone, 1996).

2. Sally Newcomb, "Contributions of British Experimentalists to the Discipline of Geology 1780-1920," *Proceedings of the American Philosophical Society* 134 (1990): 161–225; see also Sally Newcomb, "The Laboratory in a Field Science: Geology" (paper presented at the British–North American Joint Meeting of the British Society for History of Science, the Canadian Society for the History and Philosophy of Science, and the History of Science Society, Toronto, July 27, 1992). Newcomb notes that British experimental geology was closely allied with chemistry,

mineralogy, and in the case of the geological experimenter Josiah Wedgwood, the manufacture of porcelain. Hall also had read about porcelain manufacture and visited local glass foundries.

3. Newcomb, "The Laboratory in a Field Science," 1.

4. Frank Dawson Adams, *The Birth and Development of the Geological Sciences* (New York: Dover, 1938), 239.

5. Newcomb, "Contributions of British Experimentalists," 163n2.

6. Rachel Laudan, "Ideas and Organizations: The Case of the Geological Society of London," *Isis* 68 (1977): 527–38; Rachel Laudan, "The Role of Methodology in Lyell's Science," *Studies in the History and Philosophy of Science* 13 (1982): 215–49; see also C. S. Peirce, "Deduction, Induction, and Hypothesis," 1878, in *The Essential Peirce: Selected Philosophical Writings*, ed. Nathan Houser and Christian Kloesel (Bloomington: Indiana University Press, 1992), 186–99.

7. James Hall, "On the Revolutions of the Earth's Surface," *Transactions of the Royal Society of Edinburgh* 7 (1815): 139.

8. James Hall, 1805, quoted in Newcomb, "Contributions of British Experimentalists," 175n2.

9. See Hall, "On the Revolutions of the Earth's Surface," 139–211.

10. "The Formation of Mountains (A Report on the Work of Alphonse Favre)," *Nature* 19 (1878): 104.

11. Ibid., 106.

12. Ibid., 105.

13. Auguste Daubrée, *Études synthétiques de géologie expérimentale* (Paris: Dunod, 1879), 4. All unmarked translations are mine.

14. William Whewell, *History of the Inductive Sciences from the Earliest to the Present Time*, 3 vols. (London: J. W. Parker and Son, 1857), 3: 514.

15. Thomas Mellard Reade, *The Origin of Mountain Ranges Considered Experimentally, Structurally, Dynamically, and in Relation to Their Geological History* (London: Taylor and Francis, 1886), iii.

16. Ibid., 17.

17. Ibid., 27.

18. B. N. Peach et al., "Report on the recent work of the Geological Survey in the North-west Highlands of Scotland," *Quarterly Journal of the Geological Society* 44 (1888): 378–441. See also David R. Oldroyd, *The Highlands Controversy: Constructing Geological Knowledge through Fieldwork in Nineteenth-Century Britain* (Chicago: University of Chicago Press, 1990); and Rob Butler, "Mountain Building with Henry Cadell," *Geoscientist* 14 (2004): 4–11.

19. Henry Cadell, "Experimental Researches in Mountain Building," *Transactions of the Royal Society of Edinburgh* 35 (1888): 337.

20. Cadell, "Experimental Researches in Mountain Building," 339.

21. This was a point missed by many experts at that time, leading to a bitter controversy over the interpretation of the Northwest Highlands of Scotland. See Oldroyd, *The Highlands Controversy*.

22. Cadell, "Experimental Researches in Mountain Building," 347n22.

23. Bailey Willis, "Mechanics of Appalachian Structure," *Annual Report of the United States Geological Survey, 1891–1892* 13 (1893): 219.

24. Willis would later adamantly reject Alfred Wegener's use of the same argument in defense of continental drift; see Naomi Oreskes, *The Rejection of Continental Drift: Theory and Method in American Earth Science* (New York: Oxford University Press, 1999).

25. Willis, "Mechanics of Appalachian Structure," 219.

26. Ibid.

27. Ibid.

28. E. G. Mead and J. Warren, "Notes on the Mechanics of Geologic Structures," *Journal of Geology* 28 (1920): 505–23; Edward Bennett Matthews, "Progress in Structural Geology," in *Studies in Geology*, vol. 8, ed. Matthews (Baltimore: Johns Hopkins University Press, 1927), 137–61.

29. Philip H. Kuenen, "Experiments in Connection with Daly's Hypothesis on the Formation of Submarine Canyons," *Overdrukuit Leidsche Geologische Mededeelingen* 8 (1937): 327–51; Richard M. Field et al., *The Navy Princeton Expedition to the West Indies in 1932* (Washington: U.S. Government Printing Office, 1933). The latter source also cites a wealth of references.

30. That is, about one thousand meters. To get this, one would require both a much greater thickness of ice than generally assumed and a greater geographic expanse. Kuenen considered both assumptions gratuitous. Reginald A. Daly, "Origin of Submarine Canyons," *American Journal of Science* 31 (1936): 401–20.

31. Daly, "Origin of Submarine Canyons," 410.

32. Kuenen, "Experiments," 331n31.

33. Alan S. Allwardt, "The Roles of Arthur Holmes and Harry Hess in the Development of Modern Global Tectonics" (PhD diss., University of California, Santa Cruz, 1990), 69; Oreskes, *The Rejection of Continental Drift*, 245n29.

34. Philip H. Kuenen, "The Negative Isostatic Anomalies in the East Indies (with Experiments)," *Overdrukuit Leidsche Geologische Mededeelingen* 8 (1926): 170–71.

35. David T. Griggs, "A Theory of Mountain-building," *American Journal of Science* 237, no. 9 (1939): 611–650.

36. M. King Hubbert, "Theory of Scale Models as Applied to the Study of Geologic Structures," *Bulletin of the Geological Society of America* 48 (1937): 1459–520.

37. M. King Hubbert, "Strength of the Earth," *Bulletin of the American Association of Petroleum Geologists* 29 (1945): 1651.

38. Daniel Sarewitz and Roger A. Pielke Jr., "Prediction in Science and Policy," in *Prediction: Decision-Making and the Future of Nature*, ed. Sarewitz, Pielke, and Radford Byerly Jr. (Washington, D.C.: Island Press, 2000), 23–40.

39. David B. Kitts, *The Structure of Geology* (Dallas: Southern Methodist University Press, 1977).

40. See, for example, Walter Bucher, *The Deformation of the Earth's Crust* (Prince-

ton: Princeton University Press, 1933). For a discussion of the failure of Bucher's nomological project, see Oreskes, *The Rejection of Continental Drift*, 285n29.

41. J. R. Fleming, *Meteorology in America, 1800-1870* (Baltimore: Johns Hopkins University Press, 1990).

42. R. M. Friedman, *Appropriating the Weather: Vilhelm Bjerknes and the Construction of a Modern Meteorology* (Ithaca: Cornell University Press, 1989).

43. Sverre Petterssen, *Weathering the Storm: Sverre Petterssen, the D-Day Forecast, and the Rise of Modern Meteorology*, ed. J. R. Fleming (Boston: American Meteorological Society, 2001); see also Naomi Oreskes and Ronald Rainger, "Science and Security before the Atomic Bomb: The Loyalty Case of Harald U. Sverdrup," *Studies in the History and Philosophy of Modern Physics* 31 (2000): 309–69.

44. Adri de la Bruhèze, "Radiological Weapons and Radioactive Waste in the United States: Insiders' and Outsiders' Views, 1941–55," *British Journal for the History of Science* 25 (1992): 207–27; Adri de la Bruhèze, "Closing the Ranks: Definition and Stabilization of Radioactive Wastes in the U.S. Atomic Energy Commission, 1945–1960," in *Shaping Technology/Building Society: Studies in Sociotechnical Change*, ed. Wiebe E. Bijker and John Law (Cambridge: MIT Press, 1992), 140–74.

45. Although this process was repeatedly pushed back, so that the opening of a repository for waste acceptance has yet to occur.

46. Bruce C. Heezen, "The Rift in the Ocean Floor," *Scientific American* 203 (1960): 98–110; Bruce C. Heezen, "The Deep-Sea Floor," in *Continental Drift*, ed. S. K. Runcorn (New York: Academic Press, 1960), 235–88; Bruce C. Heezen, "The World Rift System," *Tectonophysics* 8 (1969): 4–6; Bruce C. Heezen, Maurice Ewing, and Marie Tharp, "The Floors of the Oceans, Part 1: The North Atlantic," *Geological Society of America Special Paper* 65 (1959): 122.

47. Bruce C. Heezen and Charles D. Hollister, "Deep Sea Current Evidence from Abyssal Sediments," *Marine Geology* 1 (1964): 141–74; Bruce C. Heezen, Charles D. Hollister, and W. F. Rudiman, "Shaping of the Continental Rise by Deep Geostrophic Contour Currents," *Science* 152 (1966): 502–8; Bruce C. Heezen and Charles D. Hollister, *The Face of the Deep* (New York: Oxford University Press, 1971).

48. Woods Hole Oceanographic Institution (Woods Hole, Mass.) MC 31, Papers of Charles D. Hollister, Proposal Index File, Sandia Laboratories contract, Document No. 13-2559 (hereafter cited as WHOI PCH).

49. Ibid.

50. Ibid.

51. Naomi Oreskes, *Science on a Mission* (Chicago: University of Chicago Press, forthcoming).

52. For example, geochemical reaction or transport models, such as EQ3/6. See T. J. Wolery, *Calculation of Chemical Equilibrium between Aqueous Solution and Minerals: The EQ3/6 Software Package* (Livermore, Calif.: Lawrence Livermore Laboratory, 1979); see also T. J. Wolery, *Chemical Modeling of Geologic Disposal*

of Nuclear Waste: Progress Report and a Perspective (Livermore, Calif.: Lawrence Livermore Laboratory, 1980).

53. Naomi Oreskes, Kristin Shrader-Frechette, and Kenneth Belitz, "Verification, Validation, and Confirmation of Numerical Models in the Earth Sciences," *Science* 263 (1994): 641–46; Naomi Oreskes, "Evaluation (Not Validation) of Quantitative Models," *Environmental Health Perspectives* 106, supp. 6 (1998): 1453–60; Naomi Oreskes and Kenneth Belitz, "Philosophical Issues in Model Assessment," in *Model Validation: Perspectives in Hydrological Science*, ed. Malcolm G. Anderson and Paul D. Bates (London: John Wiley, 2001), 23–41; Naomi Oreskes, "The Role of Quantitative Models in Science," in *Models in Ecosystem Science*, ed. Charles D. Canham, Jonathan J. Cole, and William K. Lauenroth (Princeton: Princeton University Press, 2003), 13–31.

54. See also *Prediction: Science, Decision Making, and the Future of Nature*, ed. Daniel Sarewitz, Roger A. Pielke Jr., and Radford Byerly Jr. (Washington, D.C.: Island Press, 2000).

55. Charles Rubin, quoted in Malcolm Browne, "Evidence Hints at Higher Risk of Big Quakes," *New York Times*, July 17, 1998.

Models and Simulations in Climate Change: Historical, Epistemological, Anthropological, and Political Aspects

AMY DAHAN DALMEDICO

Since the 1930s, historians and philosophers of science have usually either approached models starting from scientific theories or have taken them to be intermediary entities between theories and real objects. The notion of the model in the sciences originated in late-nineteenth-century physics with the work of James Clerk Maxwell, Ludwig Boltzmann and Pierre Duhem,[1] but it was only really formalized as a category in connection with the efforts by Ernst Mach and the positivist school to reconstruct physics, and especially with the Vienna Circle's attempt to rebuild a unified science on a logicist basis. When it emerged in the 1930s, the notion of "model" took on two well-defined meanings: the first was described as "logical" (even though it first appeared in the field of mathematics); the second emerged in the context of the empirical sciences' attempt to conceptualize the relationship between physical, economic, and social systems, on the one hand, and their formal representations, on the other. During the 1930s, models and their role in scientific knowledge also became the object of philosophical scrutiny in the context of the Vienna Circle's reflections on the verification of scientific theories (logical positivism) and their empirical falsification (Karl Popper). In all cases, theory thus provided the starting point for thinking about models, which were primarily understood as intermediary entities, transitional objects between theory and a given empirical object. Philosophers concerned themselves primarily with the relationship between models and theories, the supposed realism or representational character of models, and the purity of the methods—analogies and metaphors in particular—used for constructing them.

A look at recent scientific practice, however, especially that generated in the past few decades by computers and numerical simulations, reveals this framework to be largely inadequate for analyzing contemporary models, as an increasing number of recent historical and theoretical studies have shown. Mary Morgan and Margaret Morrison in particular have pointed to the ubiquitous presence of models (and especially of simulations) that are *both* instruments of exploration *and* autonomous objects.[2] Investigations of model systems, including several contributions in this volume, underline the shortcomings of the classical normative epistemology of models.

Historical studies have shown that even as far back as the 1950s many models did not fit the traditional conception.[3] Many examples of models can be found that were developed for very pragmatic purposes, with theory playing almost no part in the process. The Second World War and the Cold War spawned a range of multidisciplinary activities that in the United States benefited from the boom in applied mathematics, such as operations research, systems analysis, and communications engineering (including electrical, phone and servo-mechanical communications), as well as the first simulations in military nuclear research or meteorology. It seems as if postwar modeling to a large extent rejected mechanical reductionism, the hope for a unified science, and the search for universal truth. John von Neumann claimed at the time that the sciences aimed not to explain but merely to build models whose only legitimacy lay in their efficacy.[4] Modeling practices, always pulled between abstraction and application, now found themselves subjected to another set of contradictory forces: should they be first and foremost predictive and operational, or cognitive and explanatory? The tension between understanding and forecasting, between cognitive and explanatory models became a growing source of conflict—and of attempts to compromise.

From the 1980s onward, as I will show here in the paradigmatic case of contemporary climatology, models can no longer be analyzed in terms of the epistemology described above. Devised in a foundational and normative spirit, this framework proves far too restrictive to account for the construction and working of models, the diversity of modeling practices, and their significance. Computers have profoundly transformed not only modeling and simulation practices but also the way in which we conceive of models. Together with this methodological turn came a shift in the objects, phenomena, and systems under consideration, a development made possible by the computer's ability to handle increasingly complex systems, to concentrate multiple interactions and feedback, and to integrate multiple scales and temporalities.

In order to understand how scientific practices and knowledge results relate to each other, the notion of model needs to be historicized through a study of its workings and functions in different historical configurations of scientific research. It also needs to be subjected to sociological analysis: modeling activities should be reinserted into their institutional, technical, and political environments, without separating the cognitive from the social elements combining within each model. We need to pay more attention to the actors—the researchers, engineers, and users of models—and to the actions themselves. This episte-mological and methodological interest in models and simulations has led me, via the case of meteorology,[5] to the issue of global climate change in the earth system. This field's models and modeling practices illustrate perfectly the need for a new theoretical and reflexive framework.

METEOROLOGY IN THE 1950S

Meteorology provides a good point of entry into climatology. Let us examine the methodology involved in the elaboration of the first numerical prediction models, developed between 1947 and 1953 within the framework of the Meteo-rological Project. This project, closely tied to the construction of a computer, was launched at Princeton University by John von Neumann, with Jule Charney as its scientific director.[6]

The Methodology of Theory Pull

The first task Charney and his colleagues faced was the need to filter atmo-spheric phenomena of meteorological interest from other, unwanted phe-nomena (gravitation, sound-wave propagation, etc.). This process of filter-ing what Charney called "atmospheric noise" had been implicitly carried out for decades by meteorologists formulating theories, but it was brought to the foreground by the introduction of the computer. Charney claimed that out-put noise could be eliminated in either of two ways: by making sure that the input was perfectly clean, or by building a filtering system into the receiver. In 1947, Charney coined his popular electrical engineering metaphor: "The atmo-sphere is a transmitter, the computer is the receiver."[7] The computer at once appeared as a nonneutral object that not only recorded data but also selected it and thereby played an active role.

Later, Charney and his colleagues adopted a unifying approach. They in-vestigated a series of pilot problems involving increasing numbers of physi-cal, numerical, and observational factors, while simultaneously implementing algorithms that integrated the models' equations. Thus, starting from a simpli-fied model whose behaviour could be computed, they compared the model's output with the observed phenomena, and then modified the model accord-ingly—usually by adding a physical factor whose initial exclusion was thought to have caused the discrepancy between the model's output and the phenomena observed. In 1950, Charney, Ragnar Fjötoft, and von Neumann published an analysis of their equations and a summary of the numerical predictions (or rather, "post-dictions," made available after the phenomenon had already taken place) obtained over a period of twenty-four hours for a few carefully selected days.[8] The model proved satisfactory for periods when the atmosphere behaved barotropically,[9] but it was still unable to predict the formation and behavior of typhoons. The Princeton group set out to improve the model by successively investigating baroclinic models of one to seven atmospheric layers.

Subsequently, in 1952, Charney wrote that "the philosophy guiding the approach to this problem [of numerical prediction] has been to construct a hierarchy of atmospheric models of increasing complexity, the features of each successive model being determined by an analysis of the shortcomings of the previous model."[10] In the first stage of the computer era, the main trend in meteorology was arguably to complexify the simplistic physical models of atmospheric behavior. As a giant calculator, the computer made possible the treatment of increasingly complex and evolved equations to produce better descriptions of the atmosphere and therefore give hope for more precise predictions.

Charney named this approach "theory pull": the general case that a theory (in practice, a combination of several theories) describing the dynamics of atmospheric phenomena "pulls" the development of models. Several factors, however, affect the apparent purity of this methodology. First, the race-against-the-clock character of numerical weather forecasting constitutes a major constraint since the procedures of collecting observation data, feeding it into the model, processing it by the computer, and producing numerical outputs have to be carried out as quickly as possible. The improvement of forecasting has to be counterbalanced with the increased time and work required for numerically processing the model and therefore constitutes a matter of constant negotiation.

It is not the case that this hierarchy of increasingly complex physical models followed a rational logic, whereby simplified numerical prediction equations were derived from comprehensive equations expressing the laws of atmospheric movements. In fact, the algorithms for integrating models depended on computer capacity, which itself evolved constantly. Not only did computer availability vary (since Princeton's machine was still under construction, Aberdeen's ENIAC had to be used) but so did memory size, the speed of computational methods, and so on. To complicate matters, the algorithms' numeric stability (linked to the sizes of the time step and the spatial grid, respectively) proved particularly problematic and had to be dealt with following a number of strict mathematical conditions that belie any logical succession of increasingly complex models.[11]

William Aspray wrote that in the postwar period, "it may be fair to say that the computer transformed meteorology into a mathematical science."[12] What can be observed, in my opinion, is rather that the technical and numerical possibilities created by computers considerably disrupted the current hierarchy of physical models, turning meteorology not so much into a mathematical science whose methodology unfolded logically and rigorously, but into an *engineering* science, with its typical series of successive adjustments. During this whole period, computers not only played the role of giant calculators able to process an increasing amount of data and bring much-needed complexity to atmospheric models but they also acted as true "inductive machines," to use Charney's 1955 wording, capable of selecting physical hypotheses and of testing their effects.[13]

Models for Understanding or for Predicting?

Toward the end of the 1950s, when improvements in the quality of meteorological forecasting seemed to have ground to a halt, the models themselves came under debate. Scientists sought either to extend the principles of filtration or to revert to earlier hydrodynamics equations, taking advantage of the availability of more powerful computers. Both trends concerned physical models, which scientists sought to make more representative of the complexity of the phenomena under consideration. Yet other researchers, including Barry Saltzman and Edward Lorenz, began investigating new avenues based on the computer's use as an inductive machine (i.e., seeking to test hypotheses separately). They examined different kinds of models. These drastically simplified so-called laboratory models were intentionally designed not to *predict* but to *understand* particular aspects of atmospheric behavior. The use of these models and the question of what could be learned from them caused great controversy and discussion within the community of numerical meteorologists, in particular at a symposium held in Tokyo in 1960.[14]

In the context of this debate, Lorenz, a meteorologist at MIT, first used a simplified twelve-equation model that he shortly afterward further reduced to a three-dimensional model. The behavior of the model, which he exhibited in 1963, was chaotic. Did these simplified models display, to use Charney's expression, "fatal defects"? To what extent did they represent a reality on which they were only very roughly based? More generally, should the difficulty of long-range forecasting be attributed to computers, to numerical methods, to models, or to the atmosphere itself? These crucial questions remained an important preoccupation for Charney and his colleagues, reluctant to renounce the possibility of forecasting altogether. The chaotic turbulence of the atmosphere and the notorious so-called butterfly effect (a sensitivity to initial conditions that shattered the old dream of long-range forecasting) were only gradually accepted. It was the outcome of a complex scientific and disciplinary but also professional and cultural process that took place in the 1970s, involving not only meteorologists but also dynamic systems mathematicians, theoretical physicists, and many others.[15]

Another crucial topic emerged at the Tokyo symposium. In its closing session, Norbert Eliassen mentioned as a particularly interesting field of research the possibility of computing the climatology of an atmosphere, thereby opening a general program undertaken by climatologists in the 1980s. Strikingly, he outlined this program in very abstract mathematical terms: "Given a planet with specific properties, distribution of oceans and continents, elevations, insulation and so on; determine the distribution of climate."[16] Eliassen suggested not only new methods of mathematical modeling but also a new philosophy of modeling, expressing tremendous ambition for numeric modeling and an ideal of what science could supply. He further added:

> This should in principle be possible, from weather forecasting techniques, by making forecasts for a long period of time and making statistics. This may become of importance for determining the climate of previous geological periods where the surface features of the earth were different from what they are now, and it may also be of importance for determining changes in climate caused by various external and internal changes of the system, changes within the atmosphere or of the earth's surface or of solar radiation. Since mankind is all the time changing the properties of our planet, there are, of course, already artificially produced changes of climate, and one is even thinking of producing such changes deliberately. It is vitally important that we shall be able to predict the effects before we try to change the properties of the planet.[17]

Deliberately modifying the climate or the atmosphere has therefore not always been condemned. In fact, a great deal of research into the modification and control of weather was conducted in the United States from the 1940s to the 1970s: provoking rain, deflecting a hurricane, or even controlling the climate through the selective addition of carbon dioxide then constituted perfectly legitimate research topics. These were gradually abandoned between 1971 and 1973 due to American society's changing attitudes toward nature, technology, and risk in this period.[18]

CLIMATE CHANGE STUDIES

Before delving into the analysis of climate models connected to the greenhouse effect and contemporary scientific practice, I wish to underline three specific characteristics of this field. These are, to an extent, interdependent, and taken together, they reveal a set of scientific, political, and professional constraints inherent to the development of models.

Anthropic Sensitivity

"Putting the global climate system in an equation" is an ambitious undertaking, a new kind of big science, which has developed over the past twenty years in parallel with the growing power of computers. It has as its central objective the measurement of the climate's sensitivity to anthropic effects, in particular, forcing by the greenhouse effect. It is important to understand that today's climatology faces the particularly difficult methodological and epistemological problem not so much of accounting for the earth's climate, but of identifying the climate's *sensitivity* to anthropic effects in order to establish whether the greenhouse effect is intensifying. Against the backdrop of enormous natural climatic variability, the lack of a reference trajectory (what would be the climate without humans' doings?) makes the validation of models particularly problematic.

A Both Scientific and Political Field

The research is also inseparable from the emergence of the greenhouse effect on the international political scene, notably at the conferences held in Rio de Janeiro (1992), Kyoto (1997), and Buenos Aires (2002), to name only the most significant. Since 1979, various international scientific and political institutions have been created for which climatic change, in the broad sense of the term (including holes in the ozone layer, acid rain, desertification, etc.), constitutes a central concern. One might believe, in keeping with traditional epistemology, that scientific knowledge remains independent from and precedes political decisions. In reality, the scientific and political realms have been shown to evolve jointly, even in instances when internal consensus could not be achieved. Climate change studies have emerged as a key element in the transformation of the world sometimes referred to as "global governance" (in the Foucauldian sense of a tight combination of knowledge, techniques, and power). The United States' withdrawal from the Kyoto protocol must doubtless be analyzed from this geopolitical point of view and opens a new phase in this governance, although I cannot further treat the subject here.

Climate change studies were from the start constructed as a hybrid political *and* scientific field whose agenda, research programs, and hierarchy of objectives were partly determined by a geopolitical agenda. This feature has proven the source of considerable tension between the scientific logic of the various researchers, laboratories, and scientific subgroups involved and the geopolitical imperatives they were expected to serve. The different ways in which nations such as France, the United Kingdom, or the United States organize and direct research have, for instance, had direct repercussions on the kind of climate research carried out in these different countries. The varied and complex social and professional landscape of climate change studies has necessarily shaped the way in which scientific stakes and priorities have been defined.

The link between science and politics has taken the form of a peculiar institution, the Intergovernmental Panel on Climate Change, or IPCC, involving several hundred scientists. Created in 1988 by the World Meteorological Organization (WMO) and the United Nations Environment Programme (UNEP), its task is to inform governments of the state of knowledge in this area. It has progressively formed three working parties studying (1) the climate's and the natural biosphere's physical and chemical system; (2) the impact of climate change on the biosphere and on socioeconomic systems (this group also works on the adaptation and vulnerability of ecosystems); (3) the mitigation of climate change. This organization has contributed to the structure of the whole research field. Numerical modeling has emerged as the only tool allowing a quantitative projection into the future. In each IPCC group's work and reports (there were three extensive reports in 1990, 1995, and 2001, with a fourth to come in 2007), models played a crucial role as the main tools used in expertise.

Integration and Feedback Loops

However, the boundaries of the fields defined by these IPCC groups are blurred; as it gradually emerged over the past few years from the growing understanding of the different factors governing climate change, the results and objectives of each one are interdependent. What is the IPCC's methodology? Its third report (2001) charts in considerable detail its climatic projections until the end of the twenty-first century. These projections were carried out only after the completion of a preliminary and necessary stage: the determination, for the same period, of evolution scenarios of forcing agents such as greenhouse gases and aerosols, a task which comes within the remit of socioeconomists.

The socioeconomists drew up a series of possible images of the future, each of which was internally coherent (in terms of demography, type of economic development, social and technological choices).[19] These images can be static snapshots of a particular moment in the future (2050, 2070, 2100, etc.), but they can also reflect possible evolutions of states, hence the use of the more dynamic term *scenario*, to which I will return. All these images are then translated using one single global variable, the concentration of carbonic gas in the atmosphere (or the number of carbon particles per million, ppm).[20] The use of this variable enables scientists to focus on the globe's temperature, which has proven much easier to manipulate than other climatic indicators (rainfall for instance), this choice being of course very simplistic.

But given the importance of interaction and feedback, this methodology proved deceptive, as scientists have begun to realize in the past few years. When preparing medium- and long-term forecasts, the first group, which a priori focuses on the mechanism of the earth's climate system (physical, atmospheric, thermal, oceanographic processes, and so forth), cannot ignore the scenarios produced by the third group. But it must also integrate (or take into account) elements of the second group's work (e.g., impacts: How will hydrology be affected? How will vegetation and agriculture be modified during the first decades of eventual warming?). Attention must be paid throughout to the strategies developed at each stage of the process. The models used by each of the three groups must therefore be able to take the results of the two others into account, and so to operate in a loop. The longer term the prediction aims to be, the more important this feedback is.

These feedback loops have only just begun to be built. And it should be noted that as simplistic as it may be, the choice of a single variable, carbon concentration (or carbon equivalent), is convenient, while, on the contrary, the impact of climate on the economy and on humankind is local, multiple, and heterogeneous. Despite a pressing social demand for a study of the local impacts of global climate change, it remains difficult to construct regional models and to interpret the results obtained.

This issue of *integration*, the last characteristic I wish to highlight here, arises within climatic models themselves and in the question of the links between the climate and the socioeconomy. It raises significant methodological, epistemological, and socioprofessional difficulties (connected to the required interdisciplinarity and to the different logic of the actors and subdisciplinary communities). Integration is necessary, given the systems' complexity, and it shapes the way in which models are collectively constructed. This takes us back to the shift I have advocated above, from studying models to investigating modeling practices. Integration also affects the certainty and reliability of the predictions derived from models. Models leaving out this or that major interaction (that is, which do not integrate it) will be less trusted. And uncertainty is closely linked to the political acceptability of all the measures and actions that one could consider implementing in order to fight climate change.

A few words on the new term *scenario*. This keyword has appeared in the modeling jargon of many disciplines (chaos studies, mesoscopic physics, earth sciences), and especially in historical disciplines in which numerical experimentation has finally been made possible (paleoclimatology, evolution theory, embryology). Halfway between a model and a narrative, the scenario allows a smooth transition between what the model produces and the accounts describing this production. It also conveys a certain modesty about the results achieved by models. In the field discussed here, the scenario often expresses a manager's or a political decision maker's philosophy of modeling, rather than a scientist's. As a consequence, the modeling of many complex systems now privileges forecasting and expertise above a deep understanding of the phenomena under consideration.

The three characteristics of the field—anthropic sensitivity, a both scientific and political coconstruction of the field, and the importance of feedbacks and integration—are interdependent. It is because the object of study is less the climate than its sensitivity to anthropic effects that the question of integration proves so decisive. Such an interdependence renders necessary the simultaneous treatment of the different scientific, operational, technical, institutional, and political aspects that shape its development. I will begin, however, by giving an insight into climate model-building processes, which reveal an amazing complexity of practices and methodological problems.[21]

ATMOSPHERIC MODELS OF THE PLANET'S CLIMATE MACHINE

Climatic modeling sits at the intersection of two scientific traditions: an older and strictly climatological tradition using simple or conceptual models,[22] and a meteorological tradition closely linked to numerical meteorological forecasting models, one that has greatly developed in the West in the past thirty-five years.

The older, climatological tradition created the simple model of the planet's radiation balance on which claims to the effect of greenhouse gases on the climate rest. Based on John Tyndall's 1861 climatic theory of carbonic gas, Svante Arrhenius (1859–1927) had already argued in the early twentieth century that the average temperature of the surface of the earth can be calculated using the gaseous absorption (by water vapor, carbon dioxide, or methane) of the infrared radiation emitted by continental or ocean surfaces.[23] This model has a simple predictive value: if the level of carbon dioxide rises, the surface temperature rises with it.

Of course, this model is not very "realistic" in that it accounts neither for local temperature variations nor for atmospheric or oceanic energy transport. I use here the term *realistic* not in a philosophical sense (in opposition to nominalism or constructivism), but in the sense scientists tend to use it: the degree to which a model is realistic depends on how far it takes elementary processes and observed phenomena into account, or, in other words, the degree to which it represents the supposed reality. However, considering that in the case of climate change the system has never been observed, realism cannot simply be equated to conformity with the observations.

General Circulation Models

In order to obtain increasingly realistic models that take into account an increasing number of processes and are able to predict an increasing number of parameters liable to verification, climatology's meteorological tradition rapidly became hegemonic. This was due, first, to a potential structural similarity between meteorological forecasting models and climate models and, second, to computing power and the floods of available data—supplied in particular by space research.[24] This meteorological tradition first aimed at the construction of large general circulation models (GCMs), that is, three-dimensional representations or simulations of the atmosphere's movements and changes in its physical state (in other words, global circulation and related meteorological, chemical, biological, and other processes).[25] In France, Météo-France relied, as early as 1975, on a meteorological numerical forecasting model for the northern hemisphere.[26] In 1988, the organization switched to a *global* model (the whole earth) to produce its forecasts. In 1992, with the appearance of a new generation of parallel computers, this model was replaced by the Arpège model (now comprising between five hundred thousand and seven hundred thousand code lines), which can operate either in numerical forecasting mode or in climate mode, depending on what routines are activated. The two versions of the model share the same dynamic core and digital resolution, but they differ in their physical parametrizations.[27] Clearly, the methodology of meteorology is deployed here to cover the planet with a numeric grid, whereby priority is always given to the numerical solution of fluid dynamics equations, for the determination of pressure and wind fields and their variations. Climate models, however, consider the atmosphere statistically and on significantly longer timescales than meteorology does.

Although they have been identified, the difficulties linked to scale remain significant. The two (vertical and horizontal) motions of the atmosphere, for instance, require a distinction between two types of scales (of the order of

0.1 km for surface turbulence and 1 km for convection, up to 1,000 km for the synoptic scale). Further, different phenomena work on different characteristic timescales. We thus have a horizontal grid a few hundred (two hundred to five hundred) kilometers long and a vertical grid covering only between one hundred meters and ten or twenty kilometers. Every GCM (there are about twenty in the world) is divided into two parts: a dynamic and a physical part, the latter representing vertical exchange processes on a scale finer than the grid. The relationship between the two is far from straightforward, and indeed constitutes a source of methodological and disciplinary tensions.

To resolve the problem numerically and to calculate variations in the atmosphere's physical state over time and across space, it is necessary to set boundary conditions and specify an initial state (as for any mathematical problem involving partial differential equations). Setting boundary conditions is immensely difficult, especially for the ground surface, where the atmosphere interacts with oceans, great glaciers, ice fields, and vegetation cover. As for the initial state of the atmosphere, it cannot be chosen arbitrarily and is supplied by WMO models, which require lengthy data assimilation periods. Numerical forecasting is especially dependent on a precise knowledge of the initial conditions (given the chaotic behavior of the atmosphere), whereas climate models rely more crucially on knowledge of boundary conditions. The equations are solved by dynamics (as a function of time), while the source terms (pressure and wind fields, etc.) are determined by physical parametrizations, the time step being of a few minutes for dynamics and from ten to thirty minutes for physics.

The Dynamics/Physics Interface and Parametrizations

The interface between dynamics and physics in GCM is always problematic. It is a source of methodological and epistemological tension, which rests on professional and disciplinary disagreement between model makers in the tradition of numerical forecasting and scientists interested in fundamental research (in, for example, physics, thermodynamics, and turbulence studies). Indeed, any physical processes taking place on a scale much smaller than the model's grid cannot be calculated on the basis of physical laws and are thus replaced by parameters. This means that they are handled *indirectly* in the dynamic model; their climatic effect is estimated rather than actually calculated. This very complex methodology is referred to as "parametrization" and concerns little-known physical processes escaping dynamics, for example, average and deep convection, clouds, radiation, the processes taking place in boundary layers such as evaporation by the vegetation cover, and so on.

Parametrization involves several steps: (1) a physical conception of the phenomenon is taken as the starting point, to be translated into stable and effective algorithms; (2) these are tested against observation; and (3) they are validated in a climate model using simulations. Problems frequently arise, for the improvement of one aspect often results in a deterioration of the global result due to incomplete control over all the feedback and compensation mechanisms at work in each model. For instance, the introduction in one GCM of a new parametrization of tropical convection caused a catastrophe when unknown and inexplicable feedback reactions appeared.[28] Substantial difficulties can also arise from a change of resolution: a parametrization that worked well at resolution T_{21} no longer functions at resolution T_{42}.[29]

Clouds perhaps best illustrate the complexity of the dynamics/physics interface. Establishing the clouds' physical role in the radiation balance or in connection with vapor physics has proven extremely difficult, and even principles remain a matter of intense debate. To elucidate them, scientists have sometimes resorted to simple, so-called conceptual models to isolate and investigate a single fundamental issue. Thus a numerical simulation—"the radiative forcing of the nebulosity," which assumes that clouds are instantly transparent to radiation—was launched in 1997 to evaluate the feedback of clouds on the radiation balance.[30] Using such a forcing amounts to isolating one single aspect of cloud feedback and, ceteris paribus, evaluating the causal importance of this factor (in particular whether the feedback is positive or negative). Here computers are used as "inductive machines," as Charney characterized this practice forty years earlier, and can only have, at best, a heuristic function. It has failed to convince most specialists, especially physicists seeking to study physical processes closely to gain a deeper understanding of the phenomena.

Since the 1980s, two Météo-France physicists, together with American colleagues at the National Center of Atmospheric Research in Boulder, Colorado, have been working on cloud phenomena on extremely small scales (less than one kilometer).[31] They combine the direct observation of clouds, radar-Doppler data, and simulations to produce very small-scale models of convection phenomena. Through the application of averaging processes, which introduce new parametrization terms, they obtain small so-called cloud resolving models. Yet this process raises problems of scale shifting: How do these clouds interact on larger scales? How do they behave collectively? The initial purpose of this research on clouds was clearly to theorize physical processes: to find a theory of so-called squall lines, to understand how a storm creates a whirlwind, to explain the formation of coherent storm structures (e.g., those that recur every three days in Africa), and so forth. As it became increasingly involved in climate change issues, this physicist's logic had to adapt to the more operational objec-

tives of the model builders and developers. It had to focus on the parametrization of phenomena (in particular convection) in order to improve models.

Among the scientists working in the field of climate change, two ideal-
typical groups could be identified, which are opposed on different levels. The
first is interested in forcing and feedback, while the second aims to build the
most comprehensive and realistic models, able to reproduce the earth's climate.
The issue of parametrization brought out this opposition: the first group believes that while parametrizations may be improved, the main objective is to
gain a better understanding of the response to disturbances;[32] for the second
group, the top priority is to produce the most realistic representations and to
come as close as possible to physical phenomena.

The representation of clouds illustrates well the contrast between these two
approaches. All researchers know that the atmosphere is very sensitive to the
slightest variations in radiative or microphysical properties. The first group believes several decades will be needed to understand the role of clouds and that
other approaches are preferable to accomplish this titanic task. These researchers only reluctantly add new levels of complexity; they prefer using simpler
models that allow a larger number of numerical simulations and the testing of
more models. Despite their reliability and efficiency, GCMs are cumbersome,
not unlike Formula One cars, as some researchers have pointed out. For some
tasks, smaller models—2CVS—can prove more useful. For scientists of the second ideal type, fundamental research comes first; physical processes should remain the central focus and become integrated into the model. All researchers,
however, are reluctant to alter a model that works. Given the number of interactions in each climate model and the number of possible error compensations,
the addition of an extra level of complexity or the modification of a particular
parametrization can disturb all the others and cause the model to diverge in
uncontrollable ways.

The opposition between these two ideal types is not a metaphysical one. It
is rooted in particular forms of knowledge, know-how, and practice. Climate is
traditionally understood as a combination of thermodynamic (heat and convection) and dynamic (atmospheric and ocean circulation) phenomena, which can
be given different relative importance. Those who privilege thermodynamics
consider the earth as a box affected mainly by thermal fluxes. They tend to focus
on the increasing concentration of carbonic gas CO_2 and its associated thermodynamic feedback—for example, of ice fields with the influence of albedo, of
clouds with the quantity of incident and reflected radiation, or of water vapor.
Dynamic properties (of the atmosphere or the oceans) are rarely mentioned in
their thermal balance. Dynamicists, in contrast, are more cautious in assessing

and using this kind of work. Their background is usually close to numerical weather forecasting and meteorological research. They pay more attention to the presence and control of errors in models, which can decrease the realism of the simulations' response to disturbances. They find attractive the way in which, for example, El Niño is currently modeled.

Of course, these two thermodynamic/dynamicist standpoints are ideal types (which are not strictly identical with the previous ones). Every researcher or small group adopts an intermediary and specific combination of these two extremes. This combination also depends on the chosen field of research: in temperate areas, dynamics govern atmospheric phenomena and clouds evolve on a large scale, whereas in Africa and in tropical areas, dynamics is subservient to thermodynamics. In practice, scientists often combine relatively simple models with more realistic GCMs. While focusing mainly on a single model, they nevertheless constantly navigate between a plurality of models and carry out explorations and comparisons using a variety of approaches. The ideal types described here are useful as a means of identifying, beyond different idiosyncrasies, the reactions and debates that recur when new problems arise.

Before examining the field's most recent developments, it will be useful to examine its forms of institutional and political inscription and the configuration of the actors involved, even if it would be an error to simplify and to quickly identify the two institutions that I present below to the previous ideal types introduced. The picture is far more versatile and complicated.

A Few Remarks on the French Configuration

In France, two scientific institutions currently work on climate change models. The first is Météo-France, a Toulouse-based public and national organization specializing in meteorological prediction using the Arpège model. It is a centralized, operational organization that, in accordance with its remit, is less interested in fundamental research than in operational aspects. It is staffed mainly by engineers, recruited as fresh graduates from the prestigious École Polytechnique, who chose to enter the Meteorology Corps and become top-ranking civil servants. Bringing with them an excellent education in mathematics and physics, these engineers receive additional training in meteorology and modeling. In the mid-1980s, Météo-France set up a research centre, the Centre National de la Recherche en Météorologie, for investigating the problems and the models of numerical forecasting and data assimilation. Since the early 1990s, the center has also begun studying the climate, urban pollution, and seasonal forecasting. At the center, research engineers rub shoulders with development engineers. In the past decade, the Meteorology Corps has diversified its sources of recruitment and extended it to a few other *grandes écoles* (e.g., the École Normale Supérieure, École Nationale Supérieure d'Agronomie, and so on). At Météo-France, the distinction is relatively clear cut between model makers and developers (who can also work on parametrizations), on the one hand, and model users, who carry out simulations and test the validity of models, on the other.

The second organization is a federation of five laboratories (including the Laboratoire de Météorologie Dynamique, the oldest in the field), which gathered in the Institut Laplace in 1994.[33] These laboratories are all associated with the National Center of Scientific Research (CNRS), and the majority of their members are researchers, engineers, and technicians who belong to this organization. Each of these laboratories handles two or three components of a general circulation model, and each researcher frequently works on three, four, and sometimes even five research topics. Models, as we have seen, contain several layers or levels,[34] organized like computer networks: phenomenology of appearances; physical, chemical, biological, or economic theory; mathematical model; numerical model; computer model. Each element of the system is represented differently on each of these levels. Researchers at the Institut Laplace can work on several of these levels, alternating fundamental research (e.g., on convection or radiation) with the improvement of models and parametrizations, research on data, simulations, work on interfaces and couplings, code writing, and so on. Research is not organized along hierarchical lines, and researchers have very diverse motivations for doing modeling work.

The Laboratoire de Météorologie Dynamique from the Institut Laplace, which had been working since the 1970s on a homemade model of the planet (made obsolete by the appearance of new calculators), in 1990 took the strategic decision to build a new model, called LMDZ, instead of working on Météo-France's Arpège model, then under construction.[35] These researchers refused to adopt Météo-France's more organized and hierarchical working style, fearing that they would become users rather than actors in the development of the new model. They wished to retain the conditions of independent research and control over their choices. At the time, the scientific authorities did not interfere with this decision, which had the effect of delaying the team, from the point of view of operational efficiency, in the international scientific race (especially against the United Kingdom and Germany). The French climate community remembers this decision with mixed feelings. France is still one of the rare countries in Europe to work on two different atmospheric models.

This episode confirms the close and interactive link between models and communities of model builders and researchers (working in the field of climate studies). A model is shaped by the group that constructs and modifies it, while a research team is built around its investigation of a model and its understanding of the model's particular characteristics.

COUPLING WITH THE OCEAN

In the continuous evolution toward more realistic models, the atmosphere-ocean interaction rapidly emerged as a primary factor due to the all-important role played by oceans in the process of heat redistribution. Great computer programs were set up in the 1990s in the United States and in Europe to produce coupled atmosphere-ocean models. This arduous task required several years of research because the ocean's time scale is significantly larger than the atmosphere's, and because important gaps exist in the available observation data. Further, the current methodology, more or less based on the physical laws of propagation, whose mathematical expression is known (Navier Stokes equations), is more difficult to apply. In oceans, thermal processes are crucial.

Cerfacs: A Computing Laboratory in Charge of the Coupling

France was then lagging behind due to the presence of two competing atmospheric models, used respectively by Météo-France (Arpège) and by the Laboratoire de Météorologie Dynamique (LMDZ), reflecting an antagonism between the two institutions. In addition, neither group worked closely with oceanographers (who had yet to consider the issue of the global ocean). The mathematician Jacques-Louis Lions, the president of the Centre National d'Études Spatiales, put all his scientific prestige behind a structuring initiative in the field of atmosphere-ocean coupling.[36] A scientific computing laboratory, located in Toulouse and close to Météo-France, Cerfacs,[37] was put in charge of studying the question of coupling. In a tense institutional context, Cerfacs sought to develop a universal methodology that might be simultaneously applied to both atmospheric models in order to couple them with the oceanographic model.

Coupling, as the Cerfacs director Jean-Claude André stressed, is above all a reflection of the way in which two milieus exchange physical quantities, which will determine their interaction. Models and exchanges thus have to be *synchronized*. Beyond the physical understanding of coupling and of the time and space scales involved, technical choices also have to be made concerning the type of machine and communication used, as synchronization implies that fields of physical quantities (pressure, wind, quantity of movement) can be switched and interpolated from one model to another.

Cerfacs chose to retain a modular conception of modeling—that is, to preserve the identity of the models to be coupled—and created a fictive layer between the two environments, atmosphere and ocean, to encapsulate both models relative to these environments.[38] Certain variables were then selected from the first environment and sent to the intermediary layer, and the same procedure was repeated from the intermediary layer to the second environment. The first simulations were carried out four or five years later, and the coupling program was finally completed in 1997. Yet as Laurent Terray explained, the program does not allow the actual coupling to take place immediately: its interface first has to be implemented in the (two, three, or more) models to be coupled—for instance, ocean-ice coupling without any atmosphere or with a simplified atmosphere.

Controversies on Flux Adjustments

In coupling with the ocean, all little-known processes are again parametrized—they are handled indirectly by having estimated through simulations their climatic effects, rather than being directly calculated. But from the time that coupled atmosphere-ocean models started being used in the United States in the 1990s, so-called climatic drift phenomena appeared, revealing the inability of these models to reproduce a stable climate. Indeed, noncoupled models (of the atmosphere, the ocean, ice) are forced by particular boundary conditions that constitute a kind of implicit modeling of the exchanges with the outside environment. This forcing enables simulations not to stray too far from a given trajectory. When models are coupled, these constraints are lifted and degrees of freedom are added to the system, which can lead to aberrant climatologies or climatic drifts. But this discrepancy is connected to the coupling itself and lies at its interface.

Model builders accordingly focused on parametrizations and on so-called flux adjustments connected to the physical plausibility of such changes. Thanks to this empirical technique, which seems better mastered today, more realistic results can be obtained with a shorter calculation time. But the technique has caused great controversy. According to a survey conducted by sociologists in 1998–1999, scientists in the United Kingdom and the United States had split into two camps with regard to these techniques. Purists were reluctant to use such adjustments and advocated new programs of fundamental research into real physical processes; more pragmatic scientists, although they admitted the downsides of these techniques, deemed them irreplaceable, considering the urgency of the issues at stake and the demands of agencies and governments. Scientists who resorted to such adjustments tended to be connected with the IPCC or were decision makers, and they often worked on scenarios. On the other hand, fundamental research programs and a privileged integration into the academic community tended to remove the need for such adjustments. The authors of the 1999 report concluded that "the debate on flux adjustments is a prism through which one can explore the model builders' social, political and scientific presuppositions."[39] The division of work between different groups—physicists, oceanographers, mathematicians, university researchers, model builders working in large centers for atmospheric science or forecasting—could therefore generate different attitudes toward computers, toward the simulations they produce, and thus toward models.

In France, in order to determine the superficial ocean's heat balance, measurement and data collection campaigns were launched to investigate cyclogenesis (the generation of ocean depressions) and the impact of wind and the state of the sea on the quantity of movement exchange between the atmosphere and the ocean. The Météo-France team at Toulouse was searching for a universal parametrization, but as one member recently noted, "a universal parametrization in this domain cannot exist! Parametrizations are conditioned by geographical zones. There is too much complexity, some processes are dominant in one zone but not in another."[40] Parametrization in a numerical model therefore appears as a method of empirically adjusting a physical process, distinct from any theoretical representation of this process by a physical law.

Another problem lay in the fact that oceanographers involved in coupling work were only really interested in tropical regions and neglected to check the quality of the oceanographical model for the higher latitudes and for ice. Many misunderstandings existed between atmosphere specialists and oceanographers, and Cerfacs found it had first to bring the two communities closer and establish the terms of a common language in order to enable the construction of the coupling code.

From 1995 onward, thanks notably to Cerfacs, the first ocean-atmosphere couplings were carried out both with the Laboratoire de Météorologie Dynamique's atmospheric model and with Météo-France's Arpège model. The two French organizations involved in climate change models have since moved closer to each other, allowing a better agreement on objectives and strategies, more fruitful scientific exchanges, and a limited distribution of tasks.

TOWARD A GENERALIZED SCIENCE OF COUPLINGS

After the ocean-atmosphere coupling, the biosphere and the climate need to be brought into constant interaction, an undertaking that has only just begun. How to model the global hydrologic cycle, or couple bio-geo-chemical cycles with the climatic system? These are some of the innumerable questions that remain unanswered, and which have enormous implications for the reliability of models and the uncertainty of their predictions. The various phenomena linked to chemical reactions (ozone, aerosols) in the stratosphere and the troposphere also need taking into account, phenomena that, as we are beginning to grasp, play a crucial role in evaluating the extent of climate change.

Every extension, every coupling, and every inclusion of a new phenomenon requires research by specialists who need to understand each other and coordinate with others, in addition to the actual computing and technical work required to harmonize different models. In these transdisciplinary collaborations, instances of mutual incomprehension and disagreement are frequent, made worse by the fact that coupled models tend to fail in their early stages. Two such instances will be briefly evoked here.

The Introduction of Surface Models

Hervé Douville has built up his own ecological niche at Météo-France, where, for the past few years, he has been studying the representation of continental surfaces and their feedback on the atmospheric climate.[41] He focuses on the role of vegetation, soil hydrology, surface water flow, and the flow of large rivers.

Traditionally, climate models did not deal with continents. But the trend toward integration created a need for soil models, and in this field, too, parametrizations were introduced. A brief description of his project illustrates the scientists' contemporary methods and the networks they belong to. Douville built a surface model based on three available models, some of whose elements he made more complex. The first studies interactions among the soil, the biosphere, and the atmosphere and can run independently. The second deals with the flow of the planet's great fluvial basins (the Amazon or Mississippi rivers) and was created in Japan. The third focuses on ice fields. Douville, for his part, was mostly interested in soil humidity and its relation to climate warming.

Douville's model features three hydrologic layers (a surface layer, a shallow layer at root level, and a deep layer) and an additional layer of snow, as well as a water reservoir within leaves. He did not include the size of leaves in this model since, as he claims, one has to know when to stop—not an obvious decision to take. As one researcher from the Laboratoire de Météorologie Dynamique explained:

> Some of us want leaf size to be interactive with temperature; the size of the
> leaves is a very practical problem, it is important for evaporation, photo-

synthesis, water balance, etc. We try to include all this in a global model [but then] we no longer know what the effects will be. . . . We believe the most complex model possible will enable prediction; one school believes this. Others believe rather that modeling is an object in itself, a tool for understanding the system, and that the purpose of modeling is understanding.[42]

Once more, we encounter the same tension, shaped by the search for increasingly realistic models, integrating an increasing number of phenomena and feedback.

Douville's methodology consists of forcing surface models with Météo-France's atmospheric model before reincorporating the soil's humidity field in the complete model, which produces a more realistic behavior. It is therefore an indirect validation of the soil model, one established via the atmospheric model. The aim is to predict seasonal anomalies, inertia, or so-called climatic memory effects in the soil. His model forms part of a European project to compare surface models—a comparison which reveals such disparity in its results that it requires constant recourse to process physics. Indeed, as Douville notes, a notion such as absolute water content varies so much from one model to another that one wonders whether it is a physical concept at all. He also stresses the need to avoid locking oneself into a single model and to compare the output with observations. The model could be correct, but the data input could be faulty in the absence of systematic measurements of soil humidity, something impossible to obtain by teledetection or aerial photography.

In addition, climate warming and soil dryness appear to be linked, but to what extent is the feedback positive? This question constitutes an important source of uncertainty, varying from one region to another, doubtless more so for African than for Indian monsoons, where feedback acts differently. In the construction of this kind of model, systematic measuring and observation campaigns (e.g., of humidity or water balance) must be undertaken. It is useless to overly refine a model if the input's quality is low.

The introduction of soils into models of the earth system's climate machine thus came along with a whole set of new conceptions, practices, actors, and models. This forms part of the continuing movement to integrate increasingly heterogeneous elements, complex mechanisms, and feedback; but the new consideration for soils also announces a return to more local matters.

An Example of Interdisciplinary Work: The Meso-NH Code

At Météo-France, individual feelings are rarely expressed. The structure of the organization and the way in which it functions confer on it an unquestionable operational efficiency. In the mid-1990s, after the first success of the atmosphere-ocean coupling and after the development of surface models, the directors decided to develop a meso-scale research model useful to the whole climate research community for work on small to medium scales, hoping it would prove useful in the long run for local numerical forecasting.

Jean-Philippe Lafforre, a research engineer at Météo-France, coordinated the project for five years. It resulted in an atmosphere and surfaces simulation system, including an ocean section, which can operate at the user's request on scales ranging from one centimeter to seventy kilometers. This system, named Meso-NH code, is equipped with all the necessary parametrizations: it can call atmospheric chemistry solvents or be coupled with an aerosol module. Its new and essential element is the previous surface model that distinguishes vegetation cover from urban areas, ice fields, and so on.

The Meso-NH project required great engineering and management skills in order to enroll and combine different kinds of competence, to coordinate research, to harmonize the different submodels' codes, and to document everything accurately. Further, a long process of pooling the different subcommunities' concepts proved necessary. Indeed, "making sure that the words used by each group and for each model have the same meaning," Lafforre explained, "requires a great number of meetings."[43] Every project like the Meso-NH code, which presupposes, in addition to the construction of new models and the adaptation of existing models, an important coupling work is first and foremost an interdisciplinary task that requires a huge effort of *translation* between the languages of different communities.

This project illustrates the cultural differences between both institutions. A CNRS scientist striving to obtain new results and to publish in international journals would probably not have been able to devote enough time to it.[44] As a researcher at the Institut Laplace said, "the rules of today's research in France favor ten small disciplinary topics where everyone is his or her own boss rather than a single large transdisciplinary project under a single leader."[45] In short, the French landscape is structured by two poles: on the one side, the CNRS, staffed by researchers who are bright, very autonomous (scientifically), but relatively individualistic and fragmented, and, on the other side, a very well-structured and increasingly technological organization whose research is mainly oriented toward operational goals.

Social and Political Dynamics

The development of climate models is largely rooted in integration and coupling; but it is also determined by the context and social and political demand. Two aspects may be mentioned in this connection: they constitute a source of debate in the communities concerned, and they have significant consequences for the direction of future research development.

BACK TO LOCAL ISSUES? Indeed, soil models are usually constructed on the meso-scale, and they are closely associated with various social demands. The scientific community dealing with such models is heterogeneous, and increasingly so: biologists work alongside agronomists, geologists, soil scientists, and forest specialists. Given the loop mechanisms, this community largely overlaps with that which studies the impact of climate change on local or regional ecosystems.

Of course, a strong link exists between the way in which nature is represented and the way in which it is assessed. Because they work using worldwide data networks, global numerical models, and processes integrated at the planetary level, IPCC scientists do not usually deal with the climate of specific countries. They speak of the global climate system or of the earth system—an object to be explored, understood, described, and managed on a planetary scale. The pragmatic meaning of climate thus tends to be black-boxed, the focus being on global quantification. Climate models cannot exactly predict how climate in this or that area of the globe will be affected.

But the impact of climate change is not uniform, and geographical disparities are great; it is predicted, for instance, that climate warming will especially affect the higher latitudes. Closing the gap between the local and the global is now perceived as an important task. Awareness of and political decision on climate issues are clearly connected to the existence of concrete and local images of climate change; there is accordingly an increasing social need for local models on the part of governments and regions. The Institut Laplace and, even more so, Météo-France have accordingly increasingly focused on *regional simulations*.

LINKS WITH THE IPCC For scientists, the decision to test (or not) the economic and energy scenarios proposed by the IPCC on the (usually coupled) climate models they work on can signify constraining commitments and painful choices. The scientific community is divided on this issue. European or international programs have often been accused of squandering precious research time with their frequent meetings and bureaucratic procedures. More fundamentally, researchers blame the IPCC for imposing increasingly unwieldly and sophisticated models. Some scientists at the Laboratoire de Météorologie Dynamique believe that in supporting and favoring research on the integration of an increasing number of interactions, the IPCC is promoting conformism at the expense of the conceptual issues raised by modeling and the overhaul of the physical foundations of parametrizations.

Météo-France's research center, a more institutional and collective body, has in contrast decided to dedicate a decisive part of its work on climate (amounting to ten thousand hours of computing) to run simulations of its models from 2050 to 2100, this with the aim of testing IPCC scenarios and to significantly contribute to the fourth report, scheduled to appear in 2007. This institution is keen to feature in this international scientific and political showcase, considering it crucial for gaining scientific credibility.

A FEW CONCLUDING REMARKS

In conclusion, and to wrap up what has been a rather descriptive journey, I propose a few synthetic remarks. They concern, first, at an anthropological level, the universe of numerical models and the status of simulations; and second, from an epistemological perspective, the generalized methodology of model coupling.

The Universe of Simulations and Models

To check whether a given climatic model accurately simulates current climate, the methodology, similar to that of numerical meteorology, ideally consists in comparing the model's output (seasonal maps of temperature, rainfall, winds, etc.) with the reality of the observed climate. In the transition from meteorology to climatology, however, simulations acquire a stronger and more fundamental status. As noted above, models are not initialized with data drawn from actual observations. The climate data to which the models are compared are heterogeneous and have to be re-calibrated, interpreted, and corrected according to models themselves used as input for other models.[46] The determination of the atmosphere's supposed initial state is one example. So-called data assimilation models run without interruption to supply general circulation models. Climate models are "data-laden," to quote Paul N. Edwards, echoing Pierre Duhem's claim that data (and hypotheses) are "theory-laden."[47] Models thus begin working with physical constants (the quantity of solar energy hitting the atmosphere, the speed of the atmosphere's rotation on its axis, etc.) in which data is averaged or supplied by other models. Models then run until they reach a state of equilibrium, which is taken to be a climate model. This simulated climate can then be compared to observed climatological averages of the real earth. It can also be used itself to simulate paleoclimates or make projections of future tendencies. The climate recorded over the past fifty years is thus regarded by climate model makers, albeit with certain precautions, as an effective anthropic forcing experiment.

When attempting to detect climate change (against the background of natural variability), information can only be gathered from models and not directly from observations, as scientists often stress. Hence the crucial importance of a rigorous methodology of numerical modeling and a critical and lucid analysis of the relationship between model builders and their models.

The Numerical Laboratory, Reanalysis, Idealized Modeling

The scientists encountered in this article and the research investigated paid explicit attention to methodological factors. L. Terray's team at Cerfacs, which describes itself as a team of model users, has, for instance, worked to define a methodology of the *numerical laboratory*.[48] It consists in establishing an inventory of available models for addressing a given problem and the practical methods for using them; in identifying what preliminary analyses are required; in finding out what precise validation principles will be employed, and so on.

Indeed, researchers usually work not with a single model, but with a hierarchy of combined or coupled models, each with its own resolution grid, a possible zoom on specific regions, and solvers that may or may not be called, and so on. The aim is to supply scientists with an interface with different "boxes" and components, with means of easily launching simulations and visualizing results quickly, and of analyzing, comparing, and interpreting them. The aim is also to integrate a software structure on this interface to ease all operations that are time-consuming without being particularly complex and that are often repeated with small, superficial changes. In short, the team has attempted simultaneously to formalize a methodology and the technical tools to apply and validate it.

Moreover, observations are riddled with errors and the databases have to be corrected using data assimilation techniques. The teams thus reanalyze models by using models constrained by observations carried out over a long period in the past in order to obtain the best initial state to serve as basis for prediction models. Validation experiments are performed on the basis of simulations over the past fifty years (which constitute the reference trajectory), but such trajectories cannot be taken at face value: tropical regions are relatively deterministic and stable, while the middle latitudes are much more chaotic. Variability therefore does not have the same meaning in both latitudes.

The model builders are also developing precise methodologies they sometimes call "idealized modeling," a somewhat surprising locution: is not every model by essence an idealization? In their understanding, idealized modeling combines within each model a drastic simplification of a given class of phenomena or environment with more realistic phenomena or environments. The corresponding methodology attempts to formalize the constant interplay of forces between ever more realistic models and conceptual models, that is, between exploratory and predictive models. Mastering complexity and the chain from the simple to the complicated requires this constant coming and going between a diversity of models and approaches.

Scientists have repeatedly rejected the idea of constructing a single, notably European, model, which might have disrupted the balance of power in the community and given official backing to its forecasts. They all reject the idea of a unique, ideal model representing the real climate system (and able to reproduce it). Instead, a different kind of project was established, PRISM (Programme for Integrated Earth System Modelling), which aims to make the different European models compatible, and whose architecture facilitates comparisons. Numerous such cross-comparison programs exist (for GCMs, chemical models, soil models, different coupled models, paleoclimatic models, etc.) and are supported by international institutions. They result in detailed methodological protocols, regular colloquia at which they are compared, and common publications.

A Model Universe? How to Escape This Virtual World

In climate change studies, models are collective constructions involving numerous inter- and transdisciplinary collaborations. Their modular structure illustrates the collective and continuous character of this process of construction. Model builders are never isolated. Each individual one can intimately know and develop only a small part of the model and must trust colleagues for the rest of the model—which makes the model, if not a black box, then perhaps a dark gray one. Yet each model builder develops a quasi-emotional relationship to the model or the part of it he or she works on, a mixture of fascination and infinite improvement.

As all specialists know and repeat, models are necessarily more polished than reality because of the Courant-Friedrichs-Lewy mathematical conditions defining the relationship between spatial and time grids, which allow the numerical resolution of the model to converge and be meaningful. Yet scientists repeatedly point out the dangers of shutting oneself up inside the universe of one's model, of seeking to improve it indefinitely; instead, one should always go back to observations, to improve data and input. "We create a virtual world, a simulacrum of reality and we must be careful not to become its prisoner," one scientist pointed out.[49]

In short, scientists are lucid and display epistemological maturity. But the mysteries of the numerical box—feedback, error-compensation phenomena, scale interactions—inherent in the computerized treatment of such a complex system, mysteries that remain opaque even to the model builders themselves, make the plunge into the infinite intricacies of the model difficult to resist. The (still limited) return to local issues, the need for reliable forecasts and regional modeling, is bringing model builders closer to measuring and observation campaigns. This perceptible development can only help the model builders escape the temptation to *lock* themselves inside their models.[50]

Anatomy of an Antireductionist Methodology

The tension between understanding and predicting has been particularly acute in the history of meteorology. In the field of climate studies, it is also constantly expressed in scientific practices and constitutes the source of much strain, even if scientists on the whole deal with a large number of models (whether they are cognitive, predictive, conceptual, or realistic) in order to master complexity.

The search for an ever more realistic model of the earth system worries many scientists. Yet this search lies at the heart of the dominant methodology of the past fifteen years. This methodology seems capable, at least in principle, of unlimited development, with its simulation and validation procedures, its comparison protocols, and its interfacing projects. From numerical GCMs and the first (atmosphere-ocean) coupled models, it has extended to coupling generally (ice and ice fields, soils, hydrological cycles, vegetation, etc.) and to modular approaches to such couplings, as well as to their combination with chemical models. Such a development remains still partly theoretical because the methodology is limited by data availability (e.g., incomplete, inaccessible, or heterogeneous data) and by computer size. However, it seems one of the most accomplished examples of a concretely antireductionist method. Indeed, antireductionism, transdisciplinarity, and networking arguably define this methodology's anatomy.

Three new epistemological aspects finally confirm the discrepancy discussed in the introduction between the older epistemological discourse and contemporary model practices: (1) The particularly heterogeneous and disunified basis on which models of climate change are constructed. To overcome such disparities, a growing number and diversity of scientific groups form networks that exchange modules or model components, translate their notions and concepts, and learn to coordinate their coupling codes and the interface to other models. They participate in great common programs, while pursuing individual research projects. (2) Not only computers but also the World Wide Web plays a central role in practices and methodology that help to foster the interdisciplinarity it presupposes and within the professional networks that support this interdisciplinarity. Computers and the Web prove crucial to overcoming this disparity and to initiate an integrative process. (3) A shift of attention away from models and toward modeling and coupling practices is essential, for it introduces the category of actors into modeling processes. All constructed models constitute the outcome of collective work, and each model coupling represents an extension of the community of the scientists involved, as well as an increased complexity of the actors' configuration.[51]

NOTES

This essay was translated by Charlotte Bigg. The text benefited from remarks by Norton Wise and Hélène Guillemot. I thank them warmly.

1. Ludwig Boltzmann wrote a long article on models for the tenth edition of the *Encyclopaedia Britannica* (1902), in which he attributed the introduction of models in physics to Maxwell, but also to "[Hermann] Helmoltz, [Ernst] Mach, [Heinrich] Hertz and many others ([Gustav] Kirchhoff for example)." On this, see Hans Freudenthal, ed., *The Concept and the Role of the Model in Mathematics and Natural and Social Sciences* (Dordrecht, Netherlands: D. Reidel, 1961); M. Armatte and A. Dahan, "Modèles et modélisations (1950–2000): Nouvelles pratiques, nouveaux enjeux," *Revue d'Histoire des Sciences* 57 (2004): 245–303.

2. Mary S. Morgan and Margaret Morrison, *Models as Mediators: Perspectives on Natural and Social Science* (Cambridge: Cambridge University Press, 1999).

3. See Armatte and Dahan, "Modèles et modélisations."

4. John von Neumann, "Methods in the Physical Sciences," in *Collected Works*, ed. A. H. Taub (Oxford: Pergamon, 1961–63), 6:491–98.

5. See Amy Dahan Dalmedico, "History and Epistemology of Models: Meteorology (1946–1963) as a Case-Study," *Archive for History of Exact Sciences* 55 (2001): 395–422.

6. On the Meteorological Project, see William Aspray, *John von Neumann and the Origins of Modern Computing* (Cambridge: MIT Press, 1990).

7. Quoted in Aspray, *John von Neumann*, 300n73. The quote is from a letter from Charney to P. D. Thompson, dated February 12, 1967.

8. J. Charney, R. Fjörtoft, and J. von Neumann, "Numerical Integration of the Baro-
    tropic Vorticity Equation," *Tellus* 2 (1950): 237–54.

9. A model is called barotropic when pressure at any point depends solely on its
    location on the earth's surface and not on its altitude; it is called baroclinic when
    a vertical pressure component is added into the model, taking into account air
    circulation and loss of potential energy.

10. Meteorology Project Report, quoted in Dahan Dalmedico, "History and Episte-
    mology of Models," 404.

11. These are the so-called Courant-Friedrichs-Lewy conditions. See H. H. Gold-
    stine, *The Computer from Pascal to von Neumann* (Princeton: Princeton Univer-
    sity Press, 1977).

12. Aspray, *John von Neumann*, 152.

13. Charney, "Numerical Methods in Dynamic Meteorology," *Proceedings of the Na-
    tional Academy of Sciences, USA* 41 (1955): 798–802, quoted in Aspray, *John von
    Neumann*, 153.

14. For the study of these debates, see Dahan, "History and Epistemology of
    Models."

15. For a precise diachronic analysis of this story, see D. Aubin and A. Dahan, "Writ-
    ing the History of Dynamical Systems and Chaos: Longue Durée and Revolution,
    Disciplines and Cultures," *Historia Mathematica* 29 (2002): 273–339.

16. International Symposium on Numerical Weather Prediction in Tokyo, November
    1960, "Proceedings," ed. Sigekata Syōno et al. (Tokyo: Meteorological Society of
    Japan, 1962), 646.

17. Ibid.

18. See Chunglin Kwa, "The Rise and Fall of Weather Modification: Changes in
    American Attitudes towards Technology, Nature and Society," in *Changing the
    Atmosphere: Expert Knowledge and Environmental Governance*, ed. Clark A.
    Miller and Paul N. Edwards (Cambridge: MIT Press, 2001), 135–66.

19. These scenarios neither take into account the impact of climate-related general
    politics nor the emission objectives set by the Kyoto Protocol, though they do
    make allowances at different levels for governmental action, for example, the de-
    velopment of low-energy consuming technologies.

20. In the third report of 2001, four new variables were introduced, corresponding to
    the concentrations of greenhouse gases other than carbonic dioxide, referred to
    as "carbon equivalent."

21. This research deals mostly with very recent history: it is based on archival reports
    by French laboratories, as well as on anthropological investigations. Since the au-
    tumn of 2002, I have led a research project aiming at a reflexive analysis of scien-
    tific practices in the field of global climate change. Some of my conclusions are
    based on material collected by Hélène Guillemot, a PhD candidate at the Centre
    Alexandre Koyré, who is working on climate modeling in France. We also carried
    out extensive interviews with scientists in this field.

22. See R. Kandel, "Les modèles météorologiques et climatiques," in *Enquête sur la*

notion de modèle, ed. P. Nouvel (Paris: PUF, 2002), 67–98; Frederik Nebeker, *Calculating the Weather: Meteorology in the Twentieth Century* (San Diego: Academic Press, 1995). See also James Roger Fleming, ed., *Historical Essays on Meteorology, 1919-1995* (Boston: American Meteorological Society, 1996).

23. When radiative forcing takes place, warming almost always results from a decrease in terrestrial radiation toward space (greenhouse effect), whereas cooling is linked to the increased reflection of solar radiation.

24. Indeed, an important factor facilitating the emergence and shaping of this field of research was the exponentially increasing amounts of spatial observation data made available by agencies such as NASA, with its geostationary satellites and various means of observation and teledetection. Seeking to raise their public profile especially after the collapse of the Soviet Union and the end of the Cold War, these agencies began feeding laboratories and atmospheric science centers vast amounts of data. The planetary scale of space science contributed in this way to give a truly *global* character to the issue. See Paul N. Edwards, "Representing the Global Atmosphere: Computer Models, Data, and Knowledge about Climate Change," in Miller and Edwards, *Changing the Atmosphere*, 31–65.

25. Some GCMs were first tested on other planets, Mars in particular. Several climate model makers therefore come from astrophysics.

26. Météo-France is the French, Toulouse-based public organization in charge of supplying daily weather forecasting. In 1993 it was succeeded by Météorologie Nationale.

27. In addition, the compromise between physical precision and calculation costs is different for each version of the model. In climate mode, the model needs to run for years, while it runs for days only in meteorological mode. Further, the size of the numerical grid is not the same. Finally, elements such as stratospheric ozone or ocean variations, which do not affect France's weather in the short term, are left out in the forecasting model, while they have to be included in climatic models.

28. Jean-François Royer, Météo-France, interview by the author, Toulouse, January 28, 2004.

29. This means that if one changes the size of the grid, some parametrizations could not stay valid.

30. Hervé Le Treut, "Climat: Pourquoi les modèles n'ont pas tort," *La Recherche*, Mai 1997, 68–73.

31. Jean-Philippe Lafforre, interview by the author, January 29, 2004; Toulouse; Jean-Louis Reidelberger, interview by the author, January 30, 2004, Toulouse. Both physicists stayed in Boulder, Colorado, for over a year.

32. In his study of English-speaking countries, Simon Shackley calls them "seers." See Simon Shackley, "Epistemic Lifestyles in Climate Change Modeling," in Miller and Edwards, *Changing the Atmosphere*, 107–33.

33. The other laboratories include the Laboratoire des Sciences du Climat et de l'Environnement (LSCE) and the Laboratoire d'Océanographie Dynamique et de

Climatologie (LODYC), linked to the Commissariat à l'Énergie Atomique (CEA). It is the latter that constructed the ocean circulation model.

34. This is J. Y. Grandpeix's (from the Laboratoire de Météorologie Dynamique) strong wording at a workshop entitled "Modèles et systèmes complexes: le changement climatique global," Hyères, September, 14–20, 2004.

35. Fontevraud seminar, 1989. Unpublished Internal Report of the Laboratoire de Météorologie Dynamique, Archives of the Laboratory.

36. See Amy Dahan Dalmedico, *Jacques-Louis Lions, un mathématicien d'exception entre recherche, industrie et politique* (Paris: Éditions La Découverte, 2005).

37. The Cerfacs is a peculiar, hybrid structure, a private-law civil company that operates in collaboration with four large organizations: the CNRS, Météo-France, EDF (Electricité de France), and SNECMA (Société nationale d'étude et de construction de moteurs d'aviations), an aeronautics firm. The climate modeling team is a research unit associated with the CNRS. Lions was president of Cerfacs's scientific board.

38. Jean-Claude André and Laurent Terray, Cerfacs, interview by Amy Dahan and Hélène Guillemot, January 30, 2004, Toulouse.

39. Simon Shackley et al., "Adjusting to Policy Expectations in Climate Change Modeling: An Interdisciplinary Study of Flux Adjustments in Coupled Atmosphere-Ocean General Circulation Models," MIT Joint Program on the Science and Policy of Climate Change, report no. 48, 1999. See also Hélène Guillemot, (PhD diss., École des Hautes Études en Sciences Sociales, in preparation). The French survey displays a similar diversity in scientists' attitudes and standpoints. See Amy Dahan Dalmedico and Hélène Guillemot, "Changement climatique: Dynamiques scientifiques, expertise, enjeux géopolitique," *Sociologie du Travail* 48 (2006): 412–32.

40. Serge Planton, interview by the author, January 31, 2004, Toulouse.

41. Hervé Douville is an engineer trained as an agronomist who entered the Meteorology Corps in 1991. Hervé Douville, interview by Amy Dahan and Hélène Guillemot, January 28, 2004, Toulouse.

42. Interview by Hélène Guillemot, July 11, 2003.

43. Lafforre, interview.

44. Today, Météo-France is developing the AROME project, the equivalent of Meso-NH for numerical forecasting.

45. Interview by Hélène Guillemot, autumn 2003.

46. Edwards, "Representing the Global Atmosphere"; and Paul N. Edwards, "Global Climate Science, Uncertainty and Politics: Data-Laden Models, Models-Filtered Data," *Science as Culture* 8 (1999): 437–72.

47. Edwards, "Global Climate Science, Uncertainty and Politics."

48. Terray, interview.

49. Lafforre, interview.

50. For other anthropological developments, see Myanna Lahsen, "Climate Rhetoric:

Constructions of Climate Science in the Age of Environmentalism" (PhD diss., Rice University, 1998); Mayanna Lahsen, "Seductive Simulations? Uncertainty Distribution around Climate Models," *Social Studies of Science* 35 (2005): 895–922.

51. As Sergio Sismondo has written, "Models become a form of glue, simultaneously epistemic and social, that allows inquiry to go forward, by connecting the ideal and the material." Sergio Sismondo, "Editor's Introduction: Models, Simulation and Their Objects," in "Modeling and Simulation," special issue, *Science in Context* 12 (1999): 258.

The Curious Case of the Prisoner's Dilemma: Model Situation? Exemplary Narrative?

MARY S. MORGAN

DILEMMA I: THE HISTORIANS' DILEMMA: GAME OR WAR?

The Prisoner's Dilemma game is one of the classic games discussed in game theory, the study of strategic decision making in situations of conflict, which stretches between mathematics and the social sciences. Game theory was primarily developed during the late 1940s and into the 1960s at a number of research sites funded by various arms of the U.S. military establishment as part of their Cold War research. As an introduction to this article on game thinking in economics, let me present the historian's dilemma in dealing with this background in its starkest form with two quotations. The first comes from Jacob Bronowski in 1954:

> The scale of the damage of Nagasaki drained the blood from my heart then [in the autumn of 1945], and does so now when I speak of it. For three miles my road lay through a desert which man had made in a second. Now, nine years later, the hydrogen bomb is ready to dwarf this scale, and to turn each mile of destruction into ten miles, and citizens and scientists stare at one another and ask: "How did we blunder into this nightmare?"[1]

Writing in 1966, Anatol Rapoport provides something of an answer to how Bronowski's 1954 nightmare came about. At least part of the blame lay at the door of game theorizing. Rapoport, a highly respected figure in game theory, was keen to point out the extreme limitations on applications of formal game theory to strategic thinking and to warn of the dangers of placing reliance on such theory in Cold War actions. In making his warning, he suggests why game theory became so important to Cold War ways of thinking. I quote his words at length, for his arguments about the role of game theory in Cold War thinking prompt some of the claims I will make later about game theory and economic reasoning:

> If two-person game theory is an extension of rational decision theory to situations in which outcomes are controlled by two decision-makers whose interests are at least partially in conflict, then the range of appli-

cations of two-person game theory ought to be the range of such situations. It is understandable why, in the period following the close of World War II, when so much attention was paid especially in the United States, to the impending power struggle between the Communist and the non-Communist worlds, the appearance of game theory on the scientific horizon was hailed with enthusiasm and with great expectations.

People had witnessed the increasing abstruseness of the sciences geared to military applications. World War I had been called the chemists' war. World War II was called the physicists' war. Towards its final phases, World War II was rapidly becoming a mathematicians' war with cybernetic devices and electronic computers beginning to play a decisive role. It is assumed in many quarters that World War III (which many feel to be a matter-of-fact culmination of existing trends) will be truly a mathematicians' war.

Moreover, mathematics is assumed in those quarters to be not merely an appendage to physical science but also the foundation of strategy. . . .

Wars to come are imagined by the strategists to be either "limited wars" or "nuclear exchanges,"[2] both being envisaged as wars of strategy rather than of attrition. It seems that those strategists who are actively concerned with the conduct of limited war view such wars as "rational" instruments of national policy, in contrast to nuclear war which, because of its awesome destructiveness, falls outside the scope of rational policy. Those strategists, on the other hand, who are concerned with nuclear "exchanges," although noncommittal about the "rationality" of such maneuvers [sic], view the potential for waging nuclear war as bargaining leverage in international affairs. They view the use of this potential as a basis for rational diplo-military policy. . . .

One finds in the writings of contemporary strategists a deliberate striving to rehabilitate war as a normal event among civilized nations. The re-establishment of high intellectual content in military strategy doubtless serves this purpose. In my opinion, the tremendous interest aroused by game theory is in no small measure due to the climate in which the rehabilitation of war, or at least of the sophisticated power struggle, was undertaken.

It becomes, therefore, extremely tempting to those actively involved in game theory and also interested in its application potential to reply in the affirmative to the question "Is game theory useful?" Since rationality in conflict enjoys extremely high prestige in our day when "realism" and "tough-mindedness" are extolled as evidence of sophistication and maturity, game theory can indeed be sold as a useful science.[3]

At this point Rapoport goes on to outline the enormous difficulties and strong limitations that arise when one tries to apply game theory beyond tic-tac-toe and into the field of international conflict.[4]

In his argument above, Rapoport links the development of game theory with a change of thinking in which war is "rehabilitated as a normal event among civilized nations." The link comes via the authority of mathematics and the ascription of a scientifically based rationality. But while Rapoport suggests that game theory was taken up because of the "civilization" of war, it seems equally part of the process that war became acceptable because it was reinterpreted in game theory terms. Whereas game theory initially provided a mathematically formulated theory of rational action in certain situations that might be *applied* to the Cold War world, the process of using game theory to think about that world turned the relationship around: the Cold War came to be seen as a set of game situations. Whereas the Prisoner's Dilemma begins as a game that might be applied, by analogy, to the nuclear arms race, it gradually comes to the point at which we understand and interpret that race as a Prisoner's Dilemma game. No doubt we are familiar enough with this idea that the first task for the historian is to recreate how strange this is.

William Poundstone's book *Prisoner's Dilemma: John von Neumann, Game Theory, and the Puzzle of the Bomb* covers the ground between Bronowski and Rapoport above. It opens with the following story, apparently a traditional African dilemma tale: "A man was crossing a river with his wife and mother. A giraffe appeared on the opposite bank. The man drew his gun [traditional?] on the beast and the giraffe said, 'If you shoot, your mother will die. If you don't shoot, your wife will die.' What should the man do?"[5] Poundstone, sensing that talking giraffes cut little ice in 1990s America, updates the tale—but only as far as the science fiction regime appropriate perhaps for 1950s or 1960s Cold War America—by replacing the talking giraffe with some mad scientists (surely some game theorists from RAND?):

> You are supposed to imagine that the pronouncements of talking giraffes are always true. You can restate the dilemma in more Western and technological terms: you, your spouse, and your mother are kidnapped by mad scientists and placed in a room with a strange machine. All three of you are bound immobile to chairs. In front of you is a push button within reach. A machine gun looms in front of your spouse and mother,

and a menacing clock ticks away on the wall. One of the scientists announces that if you push the button the mechanism will aim the gun at
your mother and shoot her dead. If you *don't* push it within sixty seconds
it will aim and fire at your spouse.[6]

These two tales produce a feeling of incredulity in me: whereas the ethical
dilemma is surely serious, the kind of situation outlined is so bizarre that it
creates a dissonance between problem and situation. Poundstone began with
the "dilemma" aspect of his project because it enabled him to trace the par-
allels between the development of game theory and the nuclear arms race. In
that context, the Prisoner's Dilemma game, understood as a model of the Cold
War conflict, represented the dilemma of whether to cooperate to prevent a
third world war or to strike first with an H-bomb, accepting the danger that
both superpowers would destroy each other. The idea that our safety might have
hung on the correct application of a Prisoner's Dilemma game is surely something we need to find incredible: we should feel the profound distance between
the seriousness of the real problem and the analytical tool, a tool that used an
extremely narrow idea of rational behavior and fitted the world situation into
a 2 x 2 matrix.[7] And, of course, Poundstone's opening is designed precisely to
make us feel the problem, to induce incredulity into the reader by posing the
horrible dilemma within an extremely oddly constructed, almost surreal, situation and narrative.

Poundstone's treatment adds something important to Rapoport's observations. It is not just that war could be written about, analyzed, and discussed
in mathematical (high intellectual) terms but that the realm of mathematical
theory treated war and international diplomacy as a series of game situations,
to be played out according to the kind of game they most resembled. Game
theorizing of the day relied not just on mathematics but also on mathematicians and social scientists playing games. In this regard, game theory was associated not so much with the civilization of war as the reduction of war studies
to the playroom.[8]

There is not nearly so much of a dilemma for the historian of economics
in writing about the role of game theory as for the historian of international
affairs. But a certain level of dissonance needs to be present even here. Otherwise we have no power to ask how it is that economists can use the Prisoner's
Dilemma (PD) game to reason about quite complicated economic behaviors
and actions. What implications did this strange little story and its matrix of
numbers have for economics?

The history of game theory in economics has a number of features in common with the discussions above. The authority of mathematics, the ascription of rationality, and the slippage between applying game thinking to economic situations and seeing those situations as games all constitute features of the spread of game thinking in economics. At first sight, the introduction of game theorizing into economics seems pretty innocuous, for the assumptions of formal game theory are closely aligned with those of neoclassical economics. But as it turns out, game theory is associated with two deep-seated changes in economics over the past half century.

One change associated with the PD is rather well understood, if not well documented. The PD game is a classic game just because it presents a dilemma economists could not resolve within the terms of their existing theory. This little two-person dilemma game turned out to have the power to undermine some cherished beliefs of neoclassical economists, and it has surreptitiously eaten away at economists' belief in the benevolence of the invisible hand and the rationality of the individual. The other change is less remarked and understood, for game thinking has introduced a form of reasoning into economics that adds case-based reasoning onto more generalizing kinds of theorizing. In this respect also economists have come to live with the PD game, and they use it very often, but not perhaps in a way they can easily characterize.

Please expect no general history of game theory in economics here.[9] Rather, I use the case of the Prisoner's Dilemma game to discuss these two changes in ways that I hope will highlight the curiosities of what has happened.

DILEMMA 2: THE PRISONER'S DILEMMA:
COLLABORATE OR DEFECT?

The Prisoner's Dilemma game is so named because it involves a dilemma for each of the individuals playing the game. It is one of a number of simple "games" studied in the social sciences embodying not just situations of conflict in which choices must be made but situations in which the players face an element of dilemma in choosing what actions to take. Unlike many games in game theory, the PD game actually began life as a game. It was possibly first played at RAND during January 1950 following a design by Merrill Flood and Melvin Dresher.[10] In the game, each of two players (John Williams from RAND and Armen Alchian from the University of California, Los Angeles) had to choose—simultaneously and in the absence of knowledge of the other's choice—one of two actions, for which they received payoffs known to both players in advance. Howard Raiffa, working at the University of Michigan on contracts for the Office of Naval Research, was at the same time investigating a similar game and carrying out game-playing experiments with it.[11] At this stage the game had no name.

The rewards from the choices made by the two players in such a game are typically given in a so-called matrix of payoffs. By convention these are treated as utilities, and under certain usual assumptions, these can be treated as monetary payoffs. Here I provide a version of the matrix of numbers given by Duncan Luce and Raiffa from their respected early text that remains a reference source in the field:[12]

	Player B, Collaborate		Player B, Defect	
Player A, Collaborate	5	5	−4	6
Player A, Defect	6	−4	−3	−3

Here B, the column player, has outcome payoffs on the right, while A, the row player, has outcome payoffs on the left, from whatever joint choices they make. For the convenience of later discussion, I have called row 1–column 1 choices the CC ("collaborate with each other") choice and row 2–column 2 choices the DD ("defect against the other") choice.

Here I have presented the game in the form in which it most usually appears in economics writings, namely, as a particular matrix of payoffs. Many variations of the numbers may be used, and while the matrix is usually symmetric, it need not be. (For example, in that first recorded game at RAND, the matrix was not symmetric, to make it less clear what the situation was.) But this variation in numbers is misleading for the numbers used cannot be just any numbers. Even slight changes in the numbers may change the matrix to represent a different game, maybe even another dilemma game.[13]

The numbers in a PD matrix are particularly chosen, but they also follow some general rule, namely, that relations between the numbers must conform to a set of inequalities. These are important in defining a PD game matrix, for they provide a more general description of the payoffs, though, surprisingly, they are rarely given.[14] Where they are given, they appear in the following form:[15]

a) $T > R > P > S$, and
b) $2R > (T + S) > 2P$

Of course, these inequalities are defined in terms that make no sense yet, and they make no sense because so far I have not reported the text that goes with the game. This in itself offers a good pointer to the importance of the text in defining the structure of the game, and whereas the game, strangely, very often appears in economics books without the inequalities, it (almost) never appears without the text.

The text attached to the matrix is the story of two prisoners and their dilemmas. Though there is some debate about who first worked on the game matrix, all parties agree that the name and story were attached to the game by Albert Tucker, a Princeton mathematician, when he wanted to use the game in a popular lecture to psychologists.[16] The PD text, which I again quote from Luce and Raiffa's first account of the game—used by the *Oxford English Dictionary* (OED) to indicate the first written usage of the term *prisoner's dilemma*— goes as follows:[17]

> The following interpretation, known as the prisoner's dilemma, is popular: Two suspects are taken into custody and separated. The district attorney is certain that they are guilty of a specific crime, but he does not have adequate evidence to convict them at a trial. He points out to each prisoner that each has two alternatives: to confess to the crime the police are sure they have done, or not to confess. If they both do not confess, then the district attorney states he will book them on some very minor trumped-up charge such as petty larceny and illegal possession of a weapon, and they will both receive minor punishment; if they both confess they will be prosecuted, but he will recommend less than the most severe sentence; but if one confesses and the other does not, then the confessor will receive lenient treatment for turning state's evidence whereas the latter will get "the book" slapped at him.[18]

Each prisoner faces a strategic choice, but this choice poses a dilemma: should the prisoner choose to cooperate with his fellow prisoner and not confess to the police, he may end up with the rewards of a small prison term, but if his fellow prisoner does the opposite, it makes his own situation the worst it could be; or should he follow his own self-interest and confess hoping his fellow will not, thus enforcing the best outcome for himself and the worst outcome for his fellow, but with the danger that his fellow will also confess, thus leaving them both worse off. That is, the dilemma for both prisoners concerns the choice of whether to trust a fellow prisoner or not, and so whether to chance the outcome of being a Sucker (not confessing when his fellow does) or giving into the Temptation of telling tales on the fellow. Of course, each one may hope that they can both reap the Reward of cooperating with each other, but there is always the possibility that both will pay the Penalty of defection. The dilemma of whether to "collaborate" and reap the joint rewards or "defect" to one's own advantage and to the loss of one's opponent was experienced by the two players involved in that first recorded playing of the game, as we can learn from the transcripts of their ongoing personal commentary as the game was played over and over again one hundred times.[19]

The economists' analysis of this dilemma is clear cut. The economist automatically assumes that each prisoner is a rational economic actor and will seek to maximize his or her individual utility (their payoff). That is, for A (row player), it is better to play row D regardless of what B (column player) does [6 > 5 and −3 > −4]; and for B (column player) it is better to play column D [6 > 5 and −3 > −4] as well. Both prisoners will act rationally by defecting (confess to the police) and both end up with penalties of −3. Although both prisoners might do better if they could agree not to confess, the text embodies the rules of the game, which forbid any discussion between them. A more modern treatment of this text would argue that even if they could agree to collaborate, the agreement is not credible or enforceable, for the payoff structure in this matrix still provides an individual temptation to defect. Thus the outcome of CC is not an "equilibrium" solution, it is not a stable outcome, whereas the DD outcome is.

This analysis reveals the meaning of the constraints on the matrix numbers, and we can now see where the inequalities terminology comes from: T is the reward from giving into temptation and defecting when his fellow does not, R is the reward to each from cooperation with each other, P is the penalty to each when both defect, and S is the loss from being a "sucker," the player who cooperates when his fellow player defects. Our previous matrix can be reinterpreted with the inequality symbols:

	Player B, Collaborate		Player B, Defect	
Player A, Collaborate	R (5)	R (5)	S (−4)	T (6)
Player A, Defect	T (6)	S (−4)	P (−3)	P (−3)

The economists' solution to this dilemma is, at one level, straightforward: rational economic man, the model man who inhabits economic theories and makes them what they are, follows an individual utility-maximizing process which, in this case, translates into making choices that end up with both players paying a penalty rather than reaping rewards. This makes economists uncomfortable, as we can see from Luce and Raiffa's comments upon the solution to the PD game: "No, there appears to be no way around this dilemma. We do not believe that there is anything irrational or perverse about the choice of α_2 and β_2 [of D, D], and we must admit that if we were actually in this position we would make these choices."[20] There was nothing wrong with the individual choices in the context of the PD game, but clearly the outcome is "wrong" in some sense—this marks the dilemma economists face.

DILEMMA 3: THE ECONOMISTS' DILEMMA:
INDIVIDUAL RATIONALITY OR INVISIBLE HAND?

Economic theory has long assumed a "rational economic man" to lie at the heart of economic reasoning. Adam Smith's economic man was a complex mixture of propensities, preferences, talents, and motives (including self-interest). A more narrowly defined *Homo economicus* only became established for economists by John Stuart Mill in 1836.[21] Mill argued that only by adopting a thin psychological profile of individuals could economics make any progress as a science. This thinly profiled or abstracted account of economic behavior, picturing man as governed overwhelmingly by wealth-seeking self-interest, was adopted within the later classical school (including Marx), and changed little in the following neoclassical and Austrian traditions. The "rationality" tag was a twentieth-century addition, where rationality was defined as utility-maximizing behavior and being consistent in choices. From Mill onward, the abstraction of rational economic man was thought to capture the essential elements of economic behavior, though it was known to serve inadequately as a description of actual behavior, which was subject to many other impulses, economic and otherwise.[22]

During the period from the 1940s through the 1960s, the period of early development for game theory, the role of rational economic man gained greater significance for he constituted a necessary component of the nexus of general equilibrium analysis and the perfectly competitive economy that formed the focus of mathematical theorizing during those years. Economists' emphasis on such narrowly defined, hard-edged economic rationality seemed during this period to turn their favorite creature into a caricature, nowhere more evident than in game theory.

This degree of rational behavior was seen as a particular problem in the PD game. The primary expression of the prisoner's dilemma is that in theoretical analysis, the outcome of individually rational and strategic behavior in the PD game leads to an outcome that is jointly irrational. That is, by following the economists' injunction to maximize their individual gain, both prisoners end up with a worse outcome than if they had both collaborated.

But perhaps, economists argued, the result might not hold if two people played a succession of such games. It is tempting, after all, to suppose that a succession of games would ensure the cooperative outcome. Even here, the *theoretical* analysis led to the same result: the usual method of analyzing the outcome of a PD game repeated a finite number of times is one of so-called backward induction.[23] It calls for first working out what will happen on the last game: because it is the last game, there is no reason for the individual to co-operate and every reason to defect, hoping that the opponent will not and thus reaping the best outcome, as in the single-game case. Since both individuals fol-low the same rationality, the last game will result in a DD outcome. Moving back one game to the penultimate game produces the same analysis. The sequence continues back to the beginning, so the theoretical analysis results in a series of bad outcomes.[24] Rational economic man has no learning power, no power to trust; he can only choose strategically the best option in any given situation.

The theoretical results were incontrovertible, leaving the economist ex-tremely uneasy—either the rationality assumption was wrong, or the outcome equilibrium was wrong. But neither could easily be given up: too much was at stake.

On the one hand, the rational economic man was deeply embedded in the high theory of the day. Because of the position of this rational character as a central building block of modern theory, softening or broadening the rational-man portrait would undermine more than just game theory or the PD result.

On the other hand, economists rely equally on another important result in economic theory, namely, the invisible-hand outcome that individuals follow-ing their own self-interest will end up doing naturally what is best for each other. According to the argument of the invisible hand, the outcome of a game in which the players follow their own self-interest should be a good one. This invisible-hand argument forms the basis of the idea that free markets and free individuals will deliver a more efficient and better outcome than that arising from government planning.

Thus, the *theory* of what happens in the PD game leaves economists in a par-ticular nasty double dilemma. Constraining the rationality of individual play, means accepting the irrational outcome that both players are worse off than if they collaborated. But accepting that irrational outcome, means denying the benevolence of the invisible hand (and thus the efficiency of the market), while weakening the rationality of individual self-interest, means potentially under-mining the mathematical work on general equilibrium analysis and perfect competition.

Economists have made various attempts to get around these dilemmas. We can classify these ways as attempts to broaden individual rationality in order to get both prisoners to the good outcome, or as ways to broaden the invisible-hand argument to accept bad outcomes. These two ways of seeking solutions to their dilemmas with the PD game are signified by economists' placement of the apostrophe in the PD. If it is the prisoner's dilemma, the question focuses on individual rationality. If it is the prisoners' dilemma, the focus is on coordination or "social" outcomes. Most economists use the term Prisoner's Dilemma and focus on the individual problem (a rare exception being David Kreps, in both his microeconomics text and his charming book on modeling and game theory, both published in 1990[25]). But when economists write about the PD as a problem of invisible-hand outcomes and issues of coordination, they more often use the form Prisoners' Dilemma.[26]

Theoretical ways around the prisoner's dilemma depend on arguments suggesting that each individual even without an opportunity to collaborate will nevertheless rationally play the cooperative move in the first place. Perhaps people have a "disposition" to trust each other? Perhaps people follow a moral code that makes them trust each other? Perhaps people begin by trusting until that is proved wrong? On the whole, economists have not found these attempts to broaden their rationality principle very convincing,[27] leaving these kinds of speculations to those more philosophically inclined.[28]

However, *experimental* work on how the PD game is played by real people for money reveals a different picture from economists' *theoretical* accounts of how rational economic man plays the game. For example, the original experiments at RAND showed that in repeated rounds of the game some collaboration would occur. Since then, there have been many experiments with the game, in both economics and psychology.[29] The most famous of these is probably the extensive set of experiments conducted by Rapoport and Albert Chammah showing how players tended to converge to either CC or DD outcomes over a series of experiments,[30] but they also report considerable variations in outcomes. Robert Axelrod's series of tournaments reported in a 1984 book investigated various strategies or sequences of moves in playing a repeated PD game.[31] These strategies were programmed to play against each other in a computer tournament. Those that did best turned out to be rather collaborative and rather forgiving— "nice" strategies if you like. The results of experimental studies of the game hit at the heart of the theoretical results obtained for the PD. These experiments, along with other research directions, have pushed economists into exploring economic behavior more broadly, maintaining rational economic man only as a benchmark and reducing economists' concern about the individual prisoner's rationality.

If the question is concerned with the prisoners' dilemma, the solution lies not in the PD itself, but in the nature of the invisible-hand argument. Economists and philosophers have a preference for a benevolent version of the invisible hand, often dated to Adam Smith's use of the term and epitomized in this oft-quoted passage from his argument: "It is not from the benevolence of the butcher, the brewer, or the baker, that we expect our dinner, but from their regard to their own interest. We address ourselves, not to their humanity but to their self-love, and never talk to them of our own necessities but of their advantages."[32] But individual self-interest does not necessarily produce benevolent outcomes, as the PD case demonstrates so easily. Here economists would do better to turn to an earlier manifestation of the invisible-hand argument available within the tradition of political economy, namely, the one proffered by Bernard Mandeville's *Private Vices, Publick Benefits* of 1705/1724.[33] In this fable, private vices lead via an invisible-hand process to public benefits, but the participants in the economy interpret the outcome as a bad one: the invisible hand is a malevolent one after all.

It is not possible here to trace the winding route and many paths by which neoclassical economists came to modify their belief in the pervasiveness of a benevolent invisible hand. Nevertheless, I believe I am on safe ground in arguing that game theory, and particularly the classic example of the PD, functions as an exemplary narrative in making this turn. In this respect I can do no better than quote from one of my London School of Economics colleagues in teaching students the lesson drawn from using the PD case during a middle-level course in economic theory: "The first law of economics is that individuals left to follow their own self-interest will reach a mutually beneficial outcome; the second law of economics is that this won't necessarily happen."[34]

DILEMMA 4: THE COMMENTATOR'S DILEMMA: MODEL SITUATION? EXEMPLARY NARRATIVE?

When we look at the way the PD game is used in economics, as opposed to the theoretical, empirical, and experimental study of the game itself, we encounter the problem of how to characterize its usage and role. On the one hand, we find that the use of mathematical notation is inextricably bound up with narratives. On the other, we find that game theory spawns models that are useful in particular cases rather than for general use. Game thinking, reasoning using game theory, does not fit easily with modern economists' picture of their science as a mathematical discipline, producing general theories derived deductively from general principles of rational behavior.

I focus on three elements to provide a characterization and thus resolve my own dilemma about how best to describe the way the PD game is used in economics. These elements I pose as questions: What are the roles of narrative?

How do economists reason about situations? Why does game theory spawn taxonomies?

What Are the Roles of Narratives?

When I first started attending to the role of narratives in economics, I took the trouble to listen carefully to how and where exactly economists used narratives. In seminars without game theory content, narratives were employed in the way economic models were used to answer questions.[35] But in economics seminars employing game theory, the role of narrative seemed truncated—narratives filled in the middle space between a set of individually rational actors with a matrix of numbers and an equilibrium solution. The narratives gave accounts of the situations, but as stories, they remained curiously unsatisfactory, for their endings were already presupposed, and the whole problem was how to get there. Although stories that are all middle do exist, it remains odd to label them as stories.[36] Yet I think now my sense that only middles were involved was, in fact, misleading.

When we look carefully at the account of the PD game given by Luce and Raiffa and the discussion of it given earlier, we can see that it is the interpretative text, not the matrix, that *implicitly* contains economists' traditional assumptions about individual rationality (i.e., which characterize the players) and about the necessity for equilibrium outcomes or solutions (a requirement of "good" economic theories) and *explicitly* contains the rules of the game (i.e., noncollaboration, simultaneous moves, etc.). All these, taken together, characterize the game situation. Thus the interpretative text is as necessary as the matrix of payoffs to the narratives that enable economists to use the game to reason in economic terms about another case at hand.[37] We might say that the narratives embody, or take for granted and depend on, the economic assumptions in the interpretive text, which is perhaps why I initially had the sense that they served only to fill in the middle of the story. In embodying the economic assumptions, they explain why the outcome is as it is, or wherein it is problematic, and how the situation might even be resolved. The narratives translate the prisoners' situation into the economic situation—they link particulars to particulars —and "explain" how it is, for example, that two large firms can end up doing damage to each other just as the prisoners end up with the double defect outcome.

We can also see how narrative elements help in thinking about how to resolve the prisoner's dilemma in Luce and Raiffa's treatment of the PD. Their broader narrative, for example, emphasizes the economists' interpretation of the rationality assumption, namely, that "neither suspect has moral qualms about or fear of squealing."[38] They explore the possibility that the prisoners might reach the collaborative outcome if cooperation were allowed, but they immediately reject this by pointing out that this scenario goes against both the rationality assumption of each player maximizing individual returns and the equilibrium outcome that follows from that assumption taken in conjunction with the payoff matrix. (That is, the joint cooperation point cannot be an equilibrium solution because both have an incentive to defect.) But this formal issue seems less strong for Luce and Raiffa than their narrative musings about whether a binding agreement to cooperate would be broken by double-crossing, and these musings are generated by the text situation and considerations of whether the game can adequately represent that situation:

> Within the criminal context, such a "double cross" may engender serious reprisals and so it might be argued that it would not be worth while. This seems, however, to deny the utility interpretation of the given numbers [in the matrix]. If we have ignored such considerations in abstracting a game from reality, we had better include the breaking of a binding agreement as an integral aspect of an enlarged game purporting to summarize the conflict of interest. Alternatively, we may suppose that the effect of breaking a binding agreement is so disastrous that it is not considered.[39]

In other words, these ways around the outcome of the PD game lead to the respecification of the rules of the game (cooperation is possible) and/or the revision of the matrix of payoffs to reflect the changes in utility. There is no way to redefine rationality to fit the case and generate the collaborative outcome: "The hopelessness that one feels in such a game as this cannot be overcome by a play on the words 'rational' and 'irrational'; it is inherent in the situation. 'There should be a law against such games!'"[40]

Why should the government step in and "pass a law against such games"? Can the government legislate against the prisoner's dilemma game? Clearly not. But in expressing such sentiments, economists are no longer arguing about the PD game, but rather about the analogous situations to which it is applied. Governments have in the past, and still do, legislate in these kinds of situation, which serves almost as a litmus test of my argument about the way that the PD game is used, namely, that economists reason about economic situations by using games as models for them.

The identification of the model case with the real-world case is not so much a matter of formal analogy but of narrative elements that allow the economist to slip easily between the two cases. For example, Shaun Hargreaves Heap and

Yanis Varoufakis suggest that "the prisoners' dilemma arises as a problem of *trust* in every elemental economic exchange because it is rare for the delivery of a good to be perfectly synchronised with the payment for it."[41] That is, the PD game can be relevant whenever we have trade at a distance over time or space (by mail, Internet, or for future delivery). In these cases, both buyer and seller have to trust each other to deliver and not to cheat on the deal to their own advantage and the disadvantage of the other exchange party. We can thus say that we often (perhaps even daily) face such a prisoner's dilemma situation. But institutions or habits of exchange and trust, those very habits that economists since Hume have thought essential to the mechanism of the market, enable us to reach the mutually beneficial outcome rather than the mutually bad one. Here we take it for granted that our market institutions and exchange habits are backed by the law of contract, which does, in effect, legislate against the PD game outcome, namely, to curtail double-crossing.

The PD game has often been used to characterize situations of competition between large firms. In this context, it has also been commonplace to reinterpret the findings of famous past economists in such terms. For example, Kreps is one of many who use the PD game to reestablish Antoine Cournot's early nineteenth-century arguments about the behavior of rival mineral water companies:

> While the [PD] story is fanciful, the basic structure of options and payoffs that characterize this game occur over and over in economics. In this basic structure players can cooperate to greater or to lesser extent. If one player unilaterally decreases the level of her cooperation, she benefits and her rival is made worse off. Consider, for example, the case of Cournot duopolists [imagine Evian and Perrier] each (independently) choosing a quantity level to bring to the market. Typically, if one firm increases its production (which is a less cooperative strategy), its profits increase, at least for a while, and the profits of its rival decrease. But (past the monopoly level of output) if both firms increase their levels of output, both do worse.[42]

Notice how Kreps moves seamlessly from the game between two players into the competition between two Cournot firms, and how, at the same time, the game rules moved from a PD game (where no cooperation is allowed) to one of possible cooperation. As Kreps continues some pages later:

> With collusion, identical firms could each supply half the monopoly quantity, and together they would obtain the monopoly profits. But this isn't an equilibrium; if one side provides half the monopoly quantity, the

other side has the incentive to supply more. . . . This isn't identical to the prisoners' dilemma game, since there we had a strictly dominant strategy for each side. But here, as there, we have (in equilibrium) each side taking actions that leave both worse off than if they could collude.[43]

Now we see this is not quite a PD game, but the characteristic outcome that characterizes the PD equilibrium remains. This, of course, remakes the exemplary point: that the self-interest outcome is not the best one. Once again the narrative matches the situation of the PD game with the economic case and enables a smooth transition in reasoning between the two while allowing a subtle change in the game specification.

Another example brings Marx into the game theory fold with his characterization of capitalists as paying their own workers low wages to maximize their own profits while hoping that all other capitalists will pay their workers high wages, thus increasing consumption demand.[44] This of course constitutes a version of the "free-rider" problem endemic in economic situations ranging from environmental pollution to labor supply. But here, the commentators have used the narration to move the situation surreptitiously from a two-person game to an n-person PD game.

This exactly reproduces the move made by Luce and Raiffa. Having discussed the prisoner's dilemma matrix and text, they give the following "alternative interpretation" for the PD game in a familiar narrative about farmers:

> As an n-person analogy to the prisoner's dilemma, consider the case of many wheat farmers where each farmer has, as an idealization, two strategies: "restricted production" and "full production." If all farmers use restricted production the price is high and individually they fare rather well; if all use full production the price is low and individually they fare rather poorly. The strategy of a given farmer, however, does not significantly affect the price level—this is the assumption of a competitive market—so that regardless of the strategies of the other farmers, he is better off in all circumstances with full production. Thus full production dominates restricted production; yet if each acts rationally they all fare poorly.[45]

The classic PD result: the invisible hand of self-interest leads the farmers to ruin. As I have noted earlier, there is no way, within the logic of the PD game, to solve this dilemma. Again, the ways around it must be sought by altering the rules of the game or some other aspect of the game played, or by widening the notion of rationality. And once again, these ways around are driven by the situation narrative providing the interplay between game and economic case:

In practice, the equilibrium [of full production] may not occur since the farmers can, and sometimes do, enter into some form of weak collusion. In addition, a farmer does not play this game just once. Rather it is repeated each year and this introduces . . . an element of collusion. Finally, sometimes the government feels as we do, steps in, and passes a law against such games. Of course, in this analysis we have neglected the consumer. When he is included collusion may not be socially desirable even if it is desirable for the farmer.[46]

In this further narrative, the commentary has moved us from a single game to a repeated PD game, and this, as Luce and Raiffa already knew in 1957 from the experiments of the 1950s, moves us into a different game, one in which collusion is more than likely, particularly when, as in the farmers' case, the number of rounds remains unknown. But it is also, as we only learn later, a game where the so-called folk theorem (a theorem known from experience) holds, so that many equilibria may occur, so almost any outcome is possible.[47]

Narratives not only bind the PD game to the economic case and provide a means of reasoning about it but they also provide the means for probing the description of the case and hence the nature of the game, changing the latter if necessary to fit the former.

How Do Economists Reason about Situations?

The application of the PD game involves reasoning about situations. Here I make use of Karl Popper's method of "situational logic" or "situational analysis," which he characterized as *the* method of economic analysis and recommended as a methodological recipe for the social sciences in general.[48] Noretta Koertge formally described the basic elements of this recipe:[49]

1. Description of the Situation:	Agent A was in a situation of type C.
2. Analysis of the Situation:	In situations of type C, the appropriate thing to do is X.
3. Rationality Principle:	Agents always act appropriately to their situations.
4. Explanandum:	(Therefore) A did X.

The idea is that an analysis of the situation combined with a rational principle of action will define what it is logical to do in each type of situation and thus enable the social scientist to "explain" an action. Popper's 1967 discussion focuses our attention away from the rationality principle and onto the situation, the *typical* situation, as the important element in the analysis: "The theoretical social sciences operate almost always by the method of constructing *typical* situations or conditions—that is, by the method of constructing models."[50]

Wade Hands has argued that this kind of logic is purpose built for economics and exemplifies exactly how economists do argue in the standard microeconomics of, for example, situations in which firms act rationally to maximize profits or consumers do to maximize utility.[51] Such situations are defined by, for example, the specific technologies of firms or the particular preferences of individuals, but in the analysis of deciding what is appropriate in such situations, microeconomists hold to some general behavioral theories or models providing standard outcomes. These theories are not very sensitive to the specifics of the situation, so that, in effect, the differences between situations hardly matter. Perhaps this is why both Hands and Bruce Caldwell,[52] who comment on Popper's ideas for economics and use Koertge's framework, nevertheless omit the important word "type" from their formulations. If there is no *type*, all situations appear either all the same or as all different individual cases. In the former case, general theory will do. In the latter, we need the kind of explanations we find in history—purpose built for each individual case. Situational analysis offers a middle level in between full generality and complete particularity only if there are different types of situations and an account of the "appropriate" thing to do for all situations of the same type.

If anything, game thinking seems much closer to fitting Popper's and Koertge's situational analysis and logic than does standard microeconomics, for here the choices or actions are very specific to the precise rules of the game and payoffs involved. In game theory and applications, the differences between situations really bite, so that even economists' thin but definite rationality principle does not always enable them to predict the outcome with ease. Yet defining the type of situation, categorizing a situation as a particular "type" of game, enables the economist to make limited generalizations about what kind of outcome will occur using the game type as the model situation.

Characterizing the analytical method in this way enables us to recognize, if we had not already done so from the examples treated above, how applications of game theory depend on the ability to match a description of an economic situation (real or hypothesized) to the description of a *type* of game in such a way that the appropriate behavior can be defined. Appropriate, for economists, as we have already seen, means not just using a rationality principle but locating some kind of an equilibrium outcome. Indeed, while much of economic game *theory* has concerned itself with defining the natures of different equilibrium concepts in different types of game situations, *applying* game theory depends on the matching of economic situations to those game types. As we have seen

above, the plausibility of this match is explored in the narrative sequences surrounding the application. If the match seems ill fitting, if the game is not appropriately specified, then the game specification is altered. As Popper argued for
the social sciences, what economists do when they use game theory is "pack or
cram our whole theoretical effort, our whole explanatory theory, in an analysis
of the *situation*: into the *model*. . . . For in this field, the empirical explanatory
theories or hypotheses are our various models, our various situational analyses.
It is these which may be empirically more or less adequate."[53]

Popper's situational analysis offers only one of the elements I need to describe
economists' reasoning with game theory.[54] Game thinking enables economists
to maintain their thin rationality and yet give an account of what will happen in
each particular type of situation: this is the domain of situational analysis.[55] Explanatory *depth* is provided by the narratives that match the situational analysis
to events in the world. In the next section, we shall see how economists gain
explanatory *breadth* across situations through the considerable possibilities for
variation in the model situations or types.

Why Does Game Theory Spawn Taxonomies?

Game theory manages to multiply the games it studies and so to generate
more types of situation that can be characterized as games. This occurs via two
mechanisms.

On the one hand, game theory has traditionally grown by filling in the holes
in a taxonomy. Luce and Raiffa introduce the PD game in a chapter entitled
"Two-Person Non-Zero-Sum Non-co-operative Games,"[56] thus designating
the game as of a particular type in a taxonomy with six categories: two versus
n people; zero-sum versus non–zero-sum; and cooperative versus noncooperative. Of course, faced with such a taxonomy, the natural theorist will find ways
to fill the empty boxes by investigating extensions of particular games within a
certain class: extending a two-person game to an *n*-person game; extending a
game without cooperation to one with cooperation; games with one period to
finite periods to infinite periods; games with zero-sums to non-zero sums; and
so forth.

This method of extending theory produces taxonomies of games that themselves change over time. New classifications emerge out of old types; new questions generate new types. Thus the taxonomies have not just grown by adding
cells over the past fifty years but they have also changed their categories. The
categories recognized in Luce and Raiffa in 1957 had entirely changed by the
time of Drew Fudenberg and Jean Tirole's text of the 1990s.[57] In the 1990s, static
versus dynamic games with complete versus incomplete information provided

the basic four-cell taxonomy with subcategories of multistage and repeated games, games in normal or strategic versus extensive form (the form of representations, which are not fully equivalent), and so forth.

The labeling of a particular game also changes as class boundaries are revised, and these change in turn as the economists come to focus on different aspects and so analyze different features of the games. The PD game is classified in the 1991 text as a static game of complete information, and its repeated version as a dynamic game of complete information, whereas in the 1957 text, it is classified under two-person, non–zero-sum, noncooperative games, and its n-person version in an equivalent n-person cell.

On the other hand, game theory also extends by attempts to characterize particular economic situations, empirical or hypothetical ones, as game situations, and thus to type them as a particular category of game. Martin Shubik in 1953 produced a taxonomy matching economic situations to game types in a table with a grand title:[58]

GENERAL THEORY OF GAMES

Cooperative Games	Semi-Cooperative Games	Non-Cooperative Games
Duopoly, Duopsony	Oligopoly	Monopoly, Monopsony
Oligopoly	Cartel Theory	Pure Competition
Cartel Theory		Cournot,
		Bertrand Duopoly
Bilateral Monopoly		Macro-economics

More typically, new categories grow from the narratives, which as we have seen go through a process of matching the economic situation with the game situation and then exploring how and why it does not fit. When it does not fit, a new version of the game is developed with slight changes in the rules, payoffs, or information arrangements. Sometimes this revision turns out to be a different type of game, yet sometimes it is like the original type. We saw examples of this kind of reasoning with the PD game earlier. It has perhaps been most clearly evident in the industrial economics literature, where a serious tradition of using game reasoning to extend the economic theory of firm competition, and to understand the exact details of empirical cases, goes back to Shubik's seminal work of the 1950s.[59]

This integration of game theory into industrial economics by Shubik initially appeared highly effective in reexploring a number of classic results in the field of the theory of the firm (dating from the nineteenth century and from the 1930s), in extending results for those cases and in allowing for the comparison of their "solutions." Game reasoning appeared to provide a new and constructive tool of analysis in the field: it offered the possibility of the analysis of strategic decisions based on situations offered by game theory in conjunction with more traditional microeconomic theory of the firm's profit-making possibilities. The approach seemed to combine the benefits of situation-based thinking with general theories of microeconomic behavior.

But by the early 1990s, the outcome, in industrial economics at least, was found less rewarding than promised. First of all, as became clearer, there were very many possible ways of characterizing economic situations as games as each game depended on many detailed specifications. For example, Sam Peltzman wrote down a "non-exhaustive list" of twenty "questions that arise in formulating and solving game-theoretic models—questions whose answers can crucially affect results."[60] These ranged from such simple ones as how many players there are and who moves first to more difficult ones concerning the nature of the equilibria in the model. The answers to these many questions characterize the rules and institutions of either a hypothetical situation imagined in the model or an empirical situation under study. Peltzman was pessimistic that "the interminable series of special cases" generated by theorists had been of any help in analyzing empirical cases.[61] Franklin Fisher was equally sharp about the way in which theoretical cases multiplied and about how they were able to provide little reliable help for an analysis of oligopoly in his comment on the folk theorem: "Anything that one might imagine as sensible can turn out to be the answer. . . . This is a case in which theory is poverty-stricken by an embarrassment of riches."[62]

Nevertheless, Fisher did not accept that finding out about the series of special cases was pointless. He interpreted the exercise not as one of a failed general theory, but as a good example of exemplifying theory, particularly useful in thinking about cartels and oligopolies, a field in which general theories had been least effective:

> When well handled, exemplifying theory can be very illuminating indeed, suggestively revealing the possibility of certain phenomena. What such theory lacks, of course, is generality. . . . The status of the theory of oligopoly is that of exemplifying theory. We know that a lot of different things *can* happen. We do not have a full, coherent, formal theory of what *must* happen or a theory that tells us how what happens depends on well-defined, measurable variables. . . . At present, oligopoly theory consists of

a large number of stories, each one an anecdote describing what might happen in some particular situation. Such stories can be very interesting indeed. Elie Wiesel . . . has said that "God made man because He loves stories," and economists (not merely game theorists) are plainly made in the divine image in this respect.[63]

Game theories exemplify typical situations or cases, and these typical cases are used to characterize empirical situations. In this sense, both theoretical work and empirical work proceed in the same way in this field, as examples of case-based reasoning. But whereas at the end of the 1930s industrial economists had four case situations, four typical situations, in their box of exemplars (perfect competition, monopoly, and two types of imperfect competition), by the 1990s game theory had filled their box of exemplifying theories for different types of situation to overflowing. For Fisher and Peltzman, typical situations had degenerated into a series of special cases or particular stories. While Fisher found this liberating, Sutton appeared more critical. He suggested that the flexibility of game theory to capture various situations was embarrassing because "given any form of behaviour observed in the market, we are now quite likely to have on hand at least one model which 'explains' it—in the sense of deriving that form of behaviour as the outcome of individually rational decisions."[64] But rather than producing an embarrassment of riches, Sutton saw it as embarrassing for game theory. As he so bluntly stated it: "In 'explaining' everything, have we explained nothing? What do these models exclude?"[65] With every economic situation potentially matched by more than one candidate model from game theory, the possibilities of using game theory for explanations in terms of types of situation—the middle-level explanatory power of situational analysis—is lost.[66] Explanatory breadth, obtained by the development of further typical situations, and therefore cells in a taxonomy, appears to have drowned in a sea of one-off individual cases and anecdotes.

EPILOGUE: A SCIENCE WITHOUT LAWS?

This article has argued through a series of four dilemmas. It began with the strategic dilemma of the Cold War and the parallel development of dilemma games in the economics of game theory, suggesting that these also posed a dilemma for historians writing about these events. It continued by discussing the way in which a particular game, the Prisoner's Dilemma game, embodied the dilemma for individuals portrayed in economic models, a dilemma that follows from economists' assumption of individual rationality and the requirement for equilibrium outcomes. Rather than a one-off case of little interest, the Prisoner's Dilemma game turned out to have exemplary qualities.

In his 1984 book on the evolution of cooperation, Axelrod remarked on the infectious quality of the PD in social psychology: "The iterated Prisoner's Dilemma has become the *E. coli* of social psychology."[67] The PD infection was equally invasive in economics, for economists came to use the case of the PD whenever they wished to discuss particular situations in which individual rationality leads to a supposedly irrational outcome.[68] Once economists started thinking about the nasty outcomes in economics that might be described in terms of a PD game, they began to find many such PD situations in the economy. For example, Michael Carter and Rodney Maddock used the two-person game to model the inflationary outcome from the interaction between government and unions.[69] Hargreaves Heap has surveyed the ways in which the PD game can be used as a model for understanding the institutional backgrounds for macroeconomic performance.[70] It has been used to model Gresham's Law (that bad money drives out good),[71] international fishing wars, productivity problems,[72] and so forth. But, and this is the dilemma explored in the third section of my essay, in developing and using the Prisoner's Dilemma game as exemplary for many such situations across many subfields of economics, economists undercut their most general lawlike claim—namely, that the social outcome of self-interested individual actions will be a good one. As a result of the infection, economics founded a second law of economics, namely, that the first law will not necessarily hold.

The final dilemma was the problem of how to characterize game reasoning in economics given that it did not fit easily into economists' perceptions of their science as a deductive, mathematical science applying general theories. Here the dilemma was resolved by an account in which narrative, situational analysis, and types of cases were fit together to portray reasoning with game theory as a kind of case-based reasoning. Situational analysis focuses the explanatory power into accurate descriptions of typical situations—into the model—so that even with a thin notion of rational behavior, specific outcomes can be deduced from the game theory for an economic situation of that type. Narrative plays the important role of enabling the economist to check that the chosen model type matches the economic situation, that is, the economic situation is accurately described in the model. By exploring the features of that match, narrative provides a sense of explanatory depth to the specific cases discussed. Explanatory breadth is derived from the development of a full taxonomy of typical cases, so that different model situations span the various empirical situations in such a way that real-world cases can be categorized in terms of a type of case. This combination should give local explanatory power at the level of types of cases. But at least in industrial economics, economists reached the point at which each case seemed different. And since explanatory power resides in the accurate description of the model situation, not in the general but thin rationality that animates the models, game theory claims both explanatory depth and breadth yet does not achieve generalized explanations. Sutton's article title "Explaining Everything, Explaining Nothing" is indicative of a science full of models but without laws.

In this process of case-based reasoning, of arguing from model situation to real case and back again, economics has undergone the same kind of transition that we observed in discussing our first dilemma, the problem of how to characterize and discuss nuclear strategy. There, we saw that part of the "civilizing" of discussions about war during the Cold War meant that nuclear war became nuclear "exchange," a reduction to economic language; and that game theory, originally a field of strategy that might be applied to Cold War problems, became the way commentators saw the Cold War world. The models of game theory began as models of, and perhaps a guide to action for, war strategies, but then the war situations themselves became seen as games. The same happened in economics. The Prisoner's Dilemma entered the OED in 1989, and I hazard a guess that it is around this time that economists stopped seeing the PD as a model of, and/or for, strategic behavior in economic situations and began to see PD situations directly in the economy. The PD game began as the lens through which economists observed less-than-happy outcomes in the economy. But they are now the things economists see in the economy, and they apparently see them almost everywhere.

NOTES

This article was written for one of the Princeton Workshops in the History of Science titled "Model Systems, Cases, and Exemplary Narratives," February 10, 2001 (and became Research Memoranda in History and Methodology of Economics 01-7, University of Amsterdam, 2001). A section of the essay was redrafted for the INEM (International Network for Economic Method) conference at Stirling in September 2002 under the title "Game Theory Explanations in Economics: Reasoning about Model Situations," and subsequent versions were given at the History of Science Society meeting in Austin, Texas, and the Second Siena Workshop on the History of Economics, both of which took place in November 2004.

I thank the Princeton workshop organisers: Angela Creager, Elizabeth Lunbeck, and Norton Wise for their invitation and responses, and Suman Seth for his commentary. I am grateful to my London School of Economics (LSE) research assistants, Till Gruene and Gabriel Molteni, who showed incredible patience with my library requests. My thanks go also to the British Academy for funding this research, to Bruce Caldwell, Robert Leonard, John Sutton, Margaret Bray, Nancy Cartwright, Ned McClennen, and my colleagues at the University of Amsterdam and at the LSE for their helpful comments.

1. Jacob Bronowski, *The Listener*, July 1, 1954, quoted in W. Poundstone, *Prisoner's Dilemma: John von Neumann, Game Theory, and the Puzzle of the Bomb* (New York: Doubleday, 1992), 265.

2. I note also how the language of nuclear war has been "civilized" into one of "exchange." David Hume argued that trade (exchange) was a civilizing influence on nations' relationships with each other; trade would replace war. Here war has become trade.

3. Anatol Rapoport, *Two-Person Game Theory: The Essential Ideas* (Ann Arbor: University of Michigan Press, 1966), 189–91.

4. He had aired these difficulties in a more public space in A. Rapoport, "The Use and Misuse of Game Theory," *Scientific American*, December 1962, 108–18.

5. Poundstone, *Prisoner's Dilemma*, 1.

6. Ibid.

7. To provide another example from Poundstone: game theorists might use the game of Chicken (two young men driving toward each other in the middle of the road to see who gives way first) to analyze the Cuban missile crisis. But, if they were Rapoport, and retained a strong sense of how far such games were from representing individual behavior, or Martin Shubik, who retained a sense of how difficult it was to represent the world in the matrix, they would warn strongly against politicians acting on the assumption that the crisis was a simple game of Chicken. See Rapoport, *Two-Person Game Theory*; and M. Shubik, "Game Theory, Behavior, and the Paradox of the Prisoner's Dilemma: Three Solutions," *Conflict Resolution* 14 (1970): 181–93.

8. Another dilemma for the historian lies in marrying the serious context and content of game thinking and the avowed purpose of its funding with the anecdotal accounts by those involved. The tenor of the anecdotes stress the playroom aspects: groups of primarily young and male mathematicians and social scientists, at least those at RAND and Princeton, had enormous fun, devising and playing games that spilled over from work into social times. As is well known, their work was funded by the military, but the accounts give little sense of the immediacy of Cold War threats, or indeed of the fact that their activities might have anything at all to do with Cold War interests. Poundstone, *Prisoner's Dilemma*, does an excellent job of laying out these disparate elements.

9. See E. Roy Weintraub, *Toward a History of Game Theory* (Durham, N.C.: Duke University Press, 1992); Sylvia Nasar, *A Beautiful Mind: A Biography of John Forbes Nash, Jr., Winner of the Nobel Prize in Economics, 1994* (New York: Simon and Schuster, 1998); Philip Mirowski, *Machine Dreams: Economics Becomes a Cyborg Science* (Cambridge: Cambridge University Press, 2001); N. Giacoli, *Mod-*

eling Rational Agents: From Interwar Economics to Early Modern Game Theory (Cheltenham, UK: Edward Elgar, 2003); R. J. Leonard, *From Red Vienna to Santa Monica* (New York: Cambridge University Press, forthcoming). Each of these works covers various aspects of the history of game theory. See entries on the PD and game theory in *The New Palgrave: A Dictionary of Economics*, ed. John Eatwell, Murray Milgate, and Peter Newman (London: Macmillan, 1998) for practitioner histories. M. S. Morgan, "Economics," in *The Modern Social Sciences*, vol. 7 of *The Cambridge History of Science*, ed. T. Porter and D. Ross (Cambridge: Cambridge University Press, 2003), 275–305, provides a more general background to the history of economics in the twentieth century.

10. M. Flood, "Some Experimental Games," in Research Memorandum RM-789, RAND Corporation, Santa Monica, 1952.

11. H. Raiffa, "Game Theory at the University of Michigan, 1948–52," in Weintraub, *Toward a History of Game Theory*, 165–75, see especially 171–73.

12. R. Duncan Luce and Howard Raiffa, *Games and Decisions: Introduction and Critical Survey* (New York: Wiley, 1957), 95.

13. For example, the Battle of the Sexes and Chicken are two other classic two-person dilemma games usually discussed alongside the PD game.

14. The game may also be represented and described in its "extensive form" (a branching tree diagram showing the choices and payoffs), or even, in the early days, in terms of its "characteristic function" (the individual and combined maximum rewards possible).

15. For example, Eric Rasmusen, *Games and Information: An Introduction to Game Theory* (Oxford: Blackwell, 1989), 30. Also see R. Axelrod, *The Evolution of Cooperation* (New York: Basic Books, 1984), 9–10.

16. Poundstone, *Prisoner's Dilemma*, 116–18.

17. The OED entry also notes Albert Tucker's more general claim to the game's provenance on the grounds of his narrative about the prisoners. Raiffa, "Game Theory at the University of Michigan, 1948-52," 173, gives an account of the disputed provenance (with RAND) of the type of game, suggesting that it was "folk knowledge" of that time.

18. Luce and Raiffa, *Games and Decisions*, 95.

19. See Poundstone, *Prisoner's Dilemma*, 2, chap. 6.

20. Luce and Raiffa, *Games and Decisions*, 96.

21. J. S. Mill, "On the Definition of Political Economy," in *Collected Works of John Stuart Mill: Essays on Economics and Society*, ed. J. M. Robson (Toronto: University of Toronto Press, 1836).

22. See M. S. Morgan, "Economic Man as Model Man: Ideal types, Idealization and Caricatures," *Journal of the History of Economic Thought* 28, no. 1 (2006): 1–27.

23. Thought experiments—or "inductive modelling," as Rapoport termed the process in which he imagined how the possible, or hypothetical, reasoning of the individuals might result in different kinds of behavior—nevertheless led to the same conclusions. See Rapoport, *Two-Person Game Theory*, chap. 10.

24. This result does not necessarily hold for the game repeated an infinite (or perhaps unknown) number of times, but it is not clear that theory tells us this.

25. David M. Kreps, *Game Theory and Economic Modelling* (Oxford: Clarendon, 1990). Also see David M. Kreps, *A Course in Microeconomic Theory* (New York: Harvester Wheatsheaf, 1990).

26. For example, G. Tullock, "Adam Smith and the Prisoners' Dilemma," *Quarterly Journal of Economics* 100 (1985): 1073–81.

27. See, for example, Shubik, "Game Theory, Behaviour, and the Paradox of the Prisoner's Dilemma."

28. See Shaun Hargreaves Heap and Yanis Varoufakis, *Game Theory: A Critical Introduction* (London: Routledge, 1995); and R. Campbell and L. Sowden, *Paradoxes of Rationality and Cooperation* (Vancouver: University of British Columbia Press, 1985).

29. The two disciplines do not necessarily make the same inferences from the same experiments, as R. J. Leonard, "Laboratory Strife: Higgling as Experimental Science in Economics and Social Psychology," in *Higgling: Transactors and Their Markets in the History of Economics*, ed. Neil de Marchi and Mary S. Morgan (Durham, N.C.: Duke University Press, 1994), 343–69, shows in the context of experiments with bargaining games.

30. A. Rapoport and A. M. Chammah, *Prisoner's Dilemma* (Ann Arbor: University of Michigan Press, 1965).

31. Axelrod, *The Evolution of Cooperation*.

32. Adam Smith, *An Inquiry into the Nature and Causes of the Wealth of Nations*, 1776, ed. R. H. Campbell and A. S. Skinner (Oxford: Oxford University Press, 1976), Book 1, Chapter 2.

33. Bernard Mandeville, *The Fable of the Bees, or, Private Vices, Publick Benefits*, ed. F. B. Kaye (Oxford: Clarendon, 1927 [1705/1724]); E. Ullman-Margalit, "Invisible-Hand Explanations," *Synthese* 39 (1978): 263–91. Ullman-Margalit's insightful discussion of invisible-hand explanations points out that on hearing them, we should find them both plausible and surprising. The PD game, I think, fits this description, and it is often associated with invisible-hand explanations, but it does not fit the other main element of her description: that they involve aggregate outcomes.

34. Margaret Bray, "Micro-economic principles," an intermediate-level course at the London School of Economics, Fall 2000.

35. See M. S. Morgan, "Models, Stories and the Economic World," *Journal of Economic Methodology* 8 (2001): 361–84.

36. Ursula Le Guin, "It Was a Dark and Stormy Night; or, Why Are We Huddling about the Campfire?" in *On Narrative*, ed. W. J. T. Mitchell (Chicago: University of Chicago Press, 1980), 194, gives an example of a minimalist middle that she is prepared to label "a whole story," one carved in runes in Carlisle Cathedral: "Tolfink carved these runes in the stone." This might be sufficient of a story to narrate a situation and outcomes as we will see in my argument's continuation (though

it is not sufficient to be a game situation unless the runes have been misread and two people were there: Tol and Fink!).

37. In contrast to L. O. Mink, "History and Fiction as Models of Comprehension," *New Literary History* 1 (1970): 541–58, who presents scientific and narrative explanations as competing, I see them as complementary here, as in my previous discussion on the role of narrative in economics. See Morgan, "Models, Stories and the Economic World."

38. Luce and Raiffa, *Games and Decisions*, 95.

39. Ibid., 96

40. Ibid., 96–97.

41. Hargreaves Heap and Varoufakis, *Game Theory*, 149.

42. Kreps, *A Course in Microeconomic Theory*, 504.

43. Ibid., 524.

44. Hargreaves Heap and Varoufakis, *Game Theory*, 154 (they quote from Marx).

45. Luce and Raiffa, *Games and Decisions*, 97

46. Ibid.

47. See D. Fudenberg and J. Tirole, *Game Theory* (Cambridge: MIT Press, 1991).

48. Karl Popper, "The Autonomy of Sociology," 1945 in *Popper Selections*, ed. D. Miller (Princeton: Princeton University Press, 1985), 345–56.

49. N. Koertge, "The Methodological Status of Popper's Rationality Principle," *Theory and Decision* 10 (1979): 87.

50. Karl Popper, "The Rationality Principle," 1967, in Miller, *Popper Selections*, 357–58; emphasis original.

51. D. W. Hands, "Falsification, Situational Analysis and Scientific Research Programs," in *Post-Popperian Methodology of Economics: Recovering Practice*, ed. Neil de Marchi (Dordrecht, the Netherlands: Kluwer, 1992), 27–31.

52. B. J. Caldwell, "Clarifying Popper," *Journal of Economic Literature* 29 (1991): 1–33.

53. Popper, "The Rationality Principle," 359–60; emphasis original.

54. It seems more than likely that my characterization will not transfer to other social sciences. For example, Rapoport, a social scientist whose base field is psychology, defines the failure of the thin rationality to provide plausible explanatory devices in game situations as one of the most important aspects of game theory, even one of its most important achievements: game theory reveals all too clearly what social scientists do not know about how humans behave.

55. This is consistent with my earlier argument about the way narratives explore the possibilities within the model. See Morgan, "Models, Stories and the Economic World."

56. Luce and Raiffa, *Games and Decisions*, 88–113.

57. Fudenberg and Tirole, *Game Theory*.

58. M. Shubik, "The Role of Game Theory in Economics," *Kyklos* 6 (1953): 27.

59. M. Shubik, *Strategy and Market Structure* (New York: Wiley, 1959).

60. S. Peltzman, "The Handbook of Industrial Organisation: A Review Article," *Journal of Political Economy* 99 (1991): 201–7.

61. Ibid., 206.

62. F. M. Fisher, "Games Economists Play: A Noncooperative View," RAND *Journal of Economics* 20 (1989): 116.

63. Ibid., 118.

64. J. Sutton, "Explaining Everything, Explaining Nothing?" *European Economic Review* 34 (1990): 506.

65. Ibid., 507.

66. In Sutton's case, this outcome has lead to the development of his so-called class-of-models approach which, in my view, reestablishes the middle level of explanatory power, not at the level of typical cases, but at the empirical level of industry characteristics. See John Sutton, *Marshall's Tendencies: What Can Economists Know?* (Cambridge: MIT Press, 2000); and my commentary on that article, "How Models Help Economists to Know," *Economics and Philosophy* 18 (2002), 5–16.

67. Axelrod, *The Evolution of Cooperation*, 28.

68. At the point when a TV quiz show can be labeled a PD game show, as happened during the period I was writing this article, it is apparent that the PD "infection" has spread beyond the social scientists' world.

69. M. Carter and R. Maddock, "Inflation: The Invisible Foot of Macroeconomics," *Economic Record* 63 (1987): 120–28.

70. S. Hargreaves Heap, "Institutions and (Short-Run) Macroeconomic Performance," *Journal of Economic Surveys* 8 (1994): 35–56.

71. G. Selgin, "Salvaging Gresham's Law: The Good, the Bad and the Illegal," *Journal of Money, Credit and Banking* 28 (1996): 637–49.

72. H. Leibenstein, "The Prisoners' Dilemma in the Invisible Hand: An Analysis of Intrafirm Productivity," *American Economic Review* 72 (1982): 92–97.

PART 3: HUMAN SCIENCES

The Psychoanalytic Case:
Voyeurism, Ethics, and Epistemology
in Robert Stoller's *Sexual Excitement*

JOHN FORRESTER

I will set out as provisional dogmatic underpinnings the following: (1) Psychoanalysis is an impossible profession not only in the Freudian sense (it is one in which, as in education and government, "one can be sure beforehand of achieving unsatisfying results"[1]) but probably also in the Lacanian sense: it is impossible successfully to transmit psychoanalytic knowledge. The case history as a genre, so anathematized by Jacques Lacan, is both the pedagogic and institutional attempt to overcome this impossibility.

(2) The case history, whilst not unique to psychoanalysis, has a distinctive function within psychoanalysis as the privileged means for attempting to convey the unique psychoanalytic experience of *both* patient and analyst.

(3) Freud's case histories were and remain exemplary in this respect. Their analysis reveals that the process of their writing obeys the same laws of transference and countertransference as the analytic situation itself. One can show with Dora—or the Wolf Man, or with the rhetorical strategies of the auto-analytic *The Interpretation of Dreams*—how the reader is implicated in Freud's countertransference, his rhetorical mastery or lack of it. Thus the transmission of psychoanalysis via Freud's clinical writing implies the repetition—or at very least the remobilization—of the original relations of transference and counter-transference evident in the relation between patient and analyst. Psychoanalytic writing is not just writing *about* psychoanalysis; it is writing subject to the same laws and processes as the psychoanalytic situation itself. In this way psycho-analysis can never free itself of the forces it attempts to describe. As a result, from one point of view, all psychoanalytic writing is exemplary of a failure. Psychoanalytic writing fails to transmit psychoanalytic knowledge because it is always simultaneously a symptom. Psychoanalytic cases, in particular, betray the founding condition of the psychoanalytic relationship, namely, its confidentiality, and are perverse through involving a third party as alibi and (un)willing partner or spectator. What kind of human practice is it for which it is impossible to bear witness, and impossible to transmit without betrayal?

(4) On the other hand, the fact that psychoanalysis enacts the very forces it attempts to capture could be taken as a demonstration of the very facts it is

communicating. We have here a knife that cuts both ways. Either psychoanalysis is irremediably contaminated by its subject matter, with no hope of objectivity, or psychoanalysis is incessantly revealing the phenomena its literary and
conversational technology endeavors to produce, despite the fact that in this
very act it willy-nilly demonstrates a lack of mastery of its own field. Ordinary
readers of texts and ordinary folk engaged in social intercourse may on occasion bemoan the fact that there is no interpretation-free zone of human relations; similarly, psychoanalysts and their critics may bemoan the fact that there
is no transference-free zone of description of those relations. Given this aporia,
what is the best strategy for the psychoanalyst? Should she fight the good fight
for objectivity, thus depriving psychoanalysis of its own logic, pretending that it
is something other than it is? Or should she brave the skeptic and undress—as
far as she dares—in public because any other way would be to pretend that she
is not naked underneath the respectable clothes of professional everyday life,
thereby also denying that nakedness is the point of wearing clothes in the first
place?

I am now going to turn to the case history with which the rest of my essay
will be principally concerned. Its author is Robert Stoller, a psychoanalyst who
lived on Sunset Boulevard in Los Angeles. The author of several books, written
from the mid-1960s onward, his principal concerns were the psychoanalysis of
gender and sexuality—he was one of the first to introduce the concept of gender into modern English.[2] Stoller's work never led him toward essentialism in
the domain of gender and sexuality, whatever its influence on others may have
been. In his work on gender he was interested in the full array of variations on
themes of gender and sexuality: in surgical transsexualism; in every variety of
gender-crossing and perversion, particularly in fetishism and sexual activities
that defy categories, such as erotic vomiting; and thence in the generalized conditions of erotic excitement. His work in the late 1970s focused on the relations
between hostility and sadomasochism.

Somewhat in the fashion that the English analyst D. W. Winnicott always
maintained an intimate connection with pediatrics, Stoller kept his nonpsy-
choanalytic psychiatric hand in—he was also a professor of psychiatry at the
University of California, Los Angeles—inventing different models of conversation, discussion, and investigation all in the service of his relentless curiosity,
his desire to find out more about gender and erotic excitement. His work is full
of hints of the unconventional channels by which he acquired his interests and
his knowledge, allowing him to write sentences such as the following: "In the

early 1960s, a fetishistic cross-dressing informant (a transvestite) left with me several pieces of his pornography. At first uninterested, I filed them away, but I returned to them a few years later, when I had come to understand better some of the dynamics of gender identity."[3] This pattern—of experience at first not comprehended and then returned to later when he began to realize its possible meanings—constitutes a ruling trope of Stoller's work: the case history I will discuss is certainly governed by it. Let me point out in passing that it is a ruling trope of Freud's work as well: the crucial insights of psychoanalysis were all given to him in the ordinary run of things, in life and in clinical practice, but could not be made use of until something else had happened to open his eyes. We find a complex dialectic of innocence and experience at work here. But it also forms one corner of what I hope to show eventually is a complex network of concepts linking secrecy, privacy, confidentiality, revelations, exhibitionism, writing, and the communication of scientific knowledge. In this essay, I will concern myself mainly with secrecy, exhibitionism, writing, and knowledge. I will fill in the links to confidentiality, the cultural conditions of the possibility of psychoanalysis, and to privacy subsequently.

In the 1980s, Stoller worked with, and again I quote, "a group of sadomaso-chists who met to teach me about s-m."[4] He went on to conduct what he called "an urban ethnography of pornography";[5] his writings on pornography were the last to be published. His final book, *Coming Attractions: The Making of an X-Rated Video*, was cowritten with Ira Levine, who had been a member of the s-m group he met with; he had worked in the pornographic industry. Levine had not been, Stoller was careful to point out, his patient, even though Stoller gave some details about his coauthor's psychopathology. In the introduction to *Coming Attractions*, Levine reports that in 1991, just as the book was being com-pleted, Stoller's life was ended by a speeding motorist on Sunset Boulevard.

THE CASE HISTORY

The case history I will be discussing was published in 1979 and makes up a large part of the book titled *Sexual Excitement: Dynamics of Erotic Life*. Like most of Stoller's other works, it is, as he makes us well aware, both slightly scandalous and deeply committed to a serious treatment of the subject matter. Throughout his career, working at a prestigious Californian university in a period during which psychoanalysis, probably sooner than anywhere else in the United States, began its nose-dive in popularity among university scientists and within the psychiatric profession, Stoller was self-conscious about the kind of discipline psychoanalysis was and the sort of knowledge one could expect from it. He summarized his overall attitude in the 1980s as follows: "At our best, we analysts are naturalistic observers of behavior with techniques—unstable but powerful—that no one else has. That's a good start."[6] In his view, one could not expect psychoanalysis to be a science because there was no possibility of confirmation. In *Sexual Excitement*, he wrote: "Most analysts believe analysis is a science. I do not, so long as one essential is missing that is found in the disciplines accepted by others as sciences: to the extent that our data are accessible to no one else, our conclusions are not subject to confirmation."[7] This preoccupation with the standing of psychoanalytic observations led Stoller to emphasize that psychoanalytic texts require underwriting by the epistemological and ethical standing of the analyst:

> There is no way to calibrate the primary research instrument of psychoanalysis, the analyst, so the audience has no reliable way to judge the accuracy of our work. I worry that we cannot be taken seriously if we do not reveal ourselves more clearly. To do so, however, may lead to messy reporting (and confuse the readers, whose own fantasies may make them feel they are peeking in on forbidden scenes). No one but the analyst can know how much the uncertain process of fixing and editing the data renders this reporting enterprise a fiction. In regard to research, says David Shakow, "Love, cherish and respect the therapist—but for heaven's sake don't trust him."[8] Artists lie to tell the truth, and scientists tell the truth to lie. (xv–xvi)

As Stoller recommends, I wish to take up a position of benign skepticism as to the story he tells in his case history. In particular, I wish to explore the complex position of the analyst as reporter, observer, participant-observer, narrator, and as a subject of the written. There are necessarily moral and epistemological implications of the task the analyst-writer sets him- or herself in the production of a true narrative or believable story. And, as I have already indicated, these implications converge interestingly with a development from within psychoanalysis, that associated with countertransference.

Sexual Excitement is some 240 pages long, excluding the notes and apparatus. It is arranged as a triptych of parts ("Hypotheses," "Data," "Theories"), followed by a fourth part entitled "Conclusions," itself followed by appendices. Part 1, some fifty pages long, consists of two hypothetico-theoretical arguments, one entitled "Sexual Excitement," the second "Primary Femininity," by way of a summary and of a preface—it is sometimes difficult to determine which.

Part 2, "Data," contains the case history of Belle; it occupies around half of the book. In the final two parts of the book, however, it is continually referred to, with new case material being added, particularly in chapter 12, the first of the Conclusions. Once Belle has entered into the book, at the very beginning of part 2, Stoller finds it difficult to take his leave of her—there are more concluding sections and valedictory notes in this narrative than any other I can think of. It recalls the complex structure of Freud's Dora case history—which also has introductions that are extensive, footnotes that add important material seemingly as an afterthought, and a postscript that rewrites what went before. Thus both Freud's and Stoller's case histories bear witness to the work required of the author-analysts to extract themselves from their relationships, from their preoccupations, with their patients. There is one crucial passage in Stoller's book, the last two pages of chapter 12, which I will discuss at length and which is not given a separate byline in the table of contents; it is a passage called "An Addendum to the Treatment." It probably was not intentional on his part, but the equivalent word to *addendum* in German is *Nachtrag*, a term that through its adjectival form of *nachträglich* has become of supreme importance in comprehending how psychoanalytic understanding (including writing) works. As we will see, Stoller's *Nachtrag* includes an epistemological bomb truly *nachträglich* in its effects.

While I will not explore the theoretical arguments surrounding the central case history, it is important to note that Stoller's explicit purpose in the book is to propose a new theory about the nature of sexual excitement, which he states as follows: "It is hostility—the desire, overt or hidden, to harm another person—that generates and enhances sexual excitement. The absence of hostility leads to sexual indifference and boredom. The hostility of erotism is an attempt, repeated over and over, to undo childhood traumas and frustrations that threatened the development of one's masculinity or femininity. The same dynamics, though in different mixes and degrees, are found in almost everyone, those labelled perverse and those not so labelled" (6). Having provided the theory the case history is meant to illustrate, I will now—finally—come to the case history. I should apologize beforehand for the length of some of the quotes—but Stoller's writing is so striking that part of the effects I will be discussing need to be placed entire and intact before the reader.

Stoller introduces his patient as follows:

> Let us call her Belle, for that suggests how she felt she was when analysis began: old-fashioned femininity; a touch of exhibitionism; gentle masochism; a slightly addled yet refreshing innocence; soft, round, dreamy

erotism; an unbounded focus on males, romance, silken garments, flowers and bees, bosoms, bare behinds, and babies.

At the start of her analysis she was twenty-four, a quiet, intelligent, attractive, well-groomed, feminine woman, white, American, Southern, middle-class, college-educated, single, Baptist (fundamentalist sect, her mother heavy on the dramas of sin, her father bored with church). (59)

We are somewhere between Hollywood, *Playboy*, and Smalltown, Alabama. We are also, one will have noticed, immediately engaged by a question about the author: is this how he, the analyst and clinical author, viewed his patient at the beginning of the analysis? Or does this vivid and curiosity-provoking description result from his ability to write his patient's view of herself on her behalf? Eight words into the first sentence we stumble on the phrase "how she felt" when analysis began, which immediately lets us loose on the problem of psychoanalytic knowledge: whose knowledge is this? Saying the words "how she felt" has the effect of raising the question "how did he feel?" How he actually felt is of course of great interest, both theoretically and practically.

Having introduced his patient, with short cameos for the various members of her family, her dream life, her style in analysis to begin with, and the central theme of her analysis—the fear and reality of abandonment during her childhood—Stoller moves in the next chapter to the principal object of the case history: Belle's prototypic daydream, which he says, it is "my task in this book . . . to take . . . and, as Belle and I did in her analysis, separate out its components" (68).

A cruel man, The Director, a Nazi type, is directing the activity. It consists of Belle being raped by a stallion, which has been aroused to a frenzy by a mare held off at a distance beyond where Belle is placed. In a circle around the periphery stand vaguely perceived men, expressionless, masturbating while ignoring each other, the Director, and Belle. She is there for the delectation of these men, including the Director, who, although he has an erection, makes no contact with her: her function is to be forced to unbearable sexual excitement and pleasure, thereby making a fool of herself before these men. She has been enslaved in this obscene exhibition of humiliation because it creates erections in these otherwise feelingless men; they stand there in phallic, brutal indifference. All that, however, is foreplay, setting the scene. What sends her excitement up and almost immediately to orgasm as she masturbates is not this scene alone, for obviously, if it were really happening, she would experience horror, not pleasure. Rather, what excites her is the addition of some detail that exacerbates her humiliation, e.g., the horse is replaced by a disreputable, ugly old man; or her excitement makes her so wild that she is making a dreadful scene; or her palpitating genitals are spotlighted to show that she has lost control of her physiology. And, behind the scenes, a part of herself permits the excitement because it (she) knows that she, who is masturbating in the real world, is not literally the same as "she" who is the suffering woman in the story. In the story, she is humiliated; in reality, she is safe. (68)

Each of the chapters that follows—"The Underground Fantasy," "Anality," "Sadomasochism," "Exhibitionism," "Lovely"—is linked to the analysis of this daydream, an analysis that stretched over several years. Each theme is latent in the daydream and is therefore justified in the light it then throws on the daydream. And each theme is worked in the same way: there is initial obscurity, a small detail emerging after years of unprofitable analytic work. This obscurity places the analyst in a characteristically frustrated and confused position. As more material emerges, he interprets, and for months he may find himself giving the same interpretations with slight variations, until one day something entirely new emerges. All at once, Belle has shifted, and with her shift, the analyst now understands what he has been doing for all those months and what he is now doing without having intended to. The themes are woven into Belle's history and woven back into her analysis. These familiar techniques eventually allow Stoller to come good on his claims: he can point to each element of the erotic daydream and show its functioning, its history, and, of course, its slow, dreamlike manifestation in the transference and countertransference of his communications with Belle. Thus in unraveling the meaning of the daydream, he also tells us the story of Belle's inner life from infancy to adulthood, the story of her analysis, and he reveals the structure of the traumatic incidents that constitute the starting point for the daydream's power and its defensive functions.

The model for Stoller's case history and for many other pieces of analytic writing is Freud's clinical writing—in this instance, most clearly the Wolf Man, whose case history revolves around the analysis of a dream and its resolution by the narrative of the primal scene. This unexpected narrative device has become the prototype of analytic understanding: the single dream text, fantasy, or sexual act whose full analysis, which may take years, is revealed as the perfect miniature in which all a person's life and modes of relationship are depicted in coded and distorted form. The trauma constituted the original scene of psychoanalysis, and then the dream became its exemplary scene; but the most surprising and yet now commonplace locus for this single set piece is the sexual or masturbatory fantasy. And Stoller's writing of Belle's daydream, to my mind, performs this task in an elegant and classical fashion. He self-consciously refers himself to Freud (and Melanie Klein), while subtly developing their theories:

> I want to underline what analysts since Freud and Melanie Klein have endlessly shown: that our mental life is experienced in the form of fantasies. These fantasies are present as scripts—stories—whose content and function can be determined. And I want to emphasize that what we call thinking or experiencing or knowing, whether it be conscious, preconscious, or unconscious, is a tightly compacted but nonetheless separable—analyzable—weave of fantasies. What we consciously think or feel is actually the algebraic summing of many simultaneous fantasies. (xiv)

Thus Stoller emerges as a supremely classical analyst in his approach to the material: in the analysis, he sets out a complete continuum between unconscious fantasy, daydream, conscious scripts, and books, films, stories from Belle's daily life and so forth. He also proves classical in a further respect: taking the erotic daydream, he shows how it can be broken down to reveal every major theme of Belle's childhood—the pattern of abandonment by her parents when she was three to six years old; the influence of her mother's silly femininity; the crucial part played in the development of the relation of her body to the world by the woman who looked after her in her mother's long absences, the Caretaker, who was preoccupied with cleanliness to the point of frequently inflicting on the little girl those anal rapes known as enemas. He is also classical in making evident, in the gaps and slippages in his writing, the strange epistemic status of all the work done on the erotic daydream, including even its status as a stable object. This is where I will start my analysis.

Opening the first working chapter of his book with this striking quotation of Belle's daydream is bound to make the reader assume that this was a conscious daydream available to Belle and the analyst in some stable, mutually acknowledgeable form. However, Stoller's account of its analysis slowly makes clear that this was far from being the case. The analyst only belatedly learnt of its existence: "A year or so into her analysis, Belle first mentioned the daydream" (68). Prior to that date, he suspected nothing—indeed, the analysis had hardly gotten going despite Belle supplying dreams, associations, and so forth. The daydream emerged in small moments, bit by bit, as the result of other work. In one session, when Stoller noted that it was odd he had never seen her angry, she responded by remembering how upset with him she had been after the previous session, and added: "'You hate everything about me.' She thinks of a battleship, of a childhood fantasy of being peacefully underground in a silent and happy community . . . and then: 'Did I ever tell you the first sex fantasy I ever had?

There are men and women performing on a stage. They are defecating and urinating.' The hour ended shortly thereafter. (Only months later did I learn that
this was not the first erotic daydream.)" (69).

In the next hour, with Stoller encouraging Belle by saying "there was value in
detailed description" (69), he writes:

> With more embarrassment and struggle than I had seen before, she said
> there was a daydream: "It's about a horse having intercourse with me."
> Silence. I then nudged her, asking her not only to announce what the fan-
> tasy was about but to reveal its details. (The resistance of secrecy often
> takes the form of people telling what they are thinking *about* but not *what*
> they are thinking.)
>
> She continued, saying that in the story she is watching herself having
> intercourse with a horse. Then, "No, that's not quite accurate. It really isn't
> myself that's watching. It's some man, and he is watching me doing this
> with a horse. And the reason that I am doing it is because if I do this with
> the horse, it will excite the man who is watching. So I do it for his sake,
> and in that way I can prove to him how great he is and that I am willing to
> make a fool of myself, humiliate myself, in order to gratify this man." (69)

From one point of view, it is obvious that the analyst will see such material
emerge slowly over time; in its essence, psychoanalysis is a process in time—
often a great deal of time. But here we catch the emergence of the daydream
when it already forms part of a complex relationship between analyst and
patient: the daydream is *for the analyst*—the process of its discovery is visibly
part of the transference of the patient onto the analyst. We are tempted to see
"the man" for whom she is making a "fool" of herself as the analyst—and the
scene of her humiliation as a rendition of the analytic scene itself.

We are not, and I wish to insist now, we are *never* seeing things happen in
this text, in this analysis, which Stoller (and Belle) did *not* see. It is never a matter of being smarter or more perceptive than they are; never even a matter of us
having insight where they necessarily remain blind. This is Stoller's text, he was
the analyst, and he is making available to his readers the analytic process that
produced the daydream. However, this does not mean that Stoller knows what
he is doing when he is writing: to be writing, or to be an analytic patient, which
all analysts have been, means to be in a position in which one cannot see things
that other positions or roles make possible. We, as readers, necessarily see something different from patients, analysts, and authors. So what I am engaged in is
a reconstruction and commentary on the analytic process, and on the different
positions that offer different vantage points from which sexual excitement, and

specifically Belle's sexual excitement, can be viewed. More exactly, this is a deconstruction—in what I take to be the original sense of that word—whereby a text is made to reveal its own internal stresses and strains, the forces within it working at odds with each other.

Over the next weeks and months, more details of the man in Belle's daydream emerged; then details of how she is tied down—a detail connected to her experience of not being able to move in intercourse because it would make her partner ejaculate ("she converted her angry suffering into growing excitement" [70]), and Stoller offers us a historical view of the gradual development of erotic daydreams, "often only a sentence at a time. I did not get the feeling that the horse story was a central one, nor, for years, was I interested in knowing its complete form" (71). Having shown how the parts of the daydream moved only slowly to center stage of the analysis, Stoller ratifies its centrality by detailing the development of Belle's erotic daydreams from childhood on: at ages seven to nine, she imagines women whipping women, with her watching; at ages ten to twelve, there are men on a stage urinating on women to make babies, with a powerful "Queen" silently watching—this was a period during which her mother was famously involved with glamorous men. Belle first appears in the scene as a girl watching the Queen: the watching made for the exciting part. By age thirteen, it is Belle herself who is the victim—those humiliating her are Amazons and their Queen; and a few weeks later, men are introduced, solely for their penises. Finally the Director replaced the Queen. With this account of the antecedents, and their place in Belle's overall development, Stoller confirms the centrality of the erotic daydream as a summary of Belle's erotic life and, as the upshot, a kind of curriculum vitae of her past erotic experience.

Having reassured his reader that all past daydreams have led to this one, Stoller also reassures a smaller class of readers, the psychoanalysts, that the daydream is sufficiently rich to count as an epitome of Belle's inner life. He lists all the themes that found expression in it: her heterosexuality; her homosexuality, via the voluptuous women conspicuous in its earlier drafts; her anality, as found in the theme of the "peaceful interior"; the central theme of her abandonment; both her masochism and her sadism; her femininity and sense of femaleness, the issues of gender with which he was strongly associated; Oedipal rivalry and primal scene, particularly with the stallion and mare. He also notes that "in early childhood, daydreams filled in for her parents' abandonments. (Before age six, when she thinks her daydreaming began, her favorite fairy tale was about a beautiful Snow Queen who lived far in the north and had a gnome working for her—not an ideal oedipal configuration.)" (67). Finally there is the question of her guilt and redemption—Stoller notes that her "heroic suffering redeems" her "being oversexed" (72). He also includes among these principal psychodynamic themes that of her audiences: "one—portrayed—that, in watching her, is the agent of her humiliation, and the other—implied—that stands witness to her martyrdom" (72). I will return to this particular theme.

Thus the daydream plausibly functions historically, structurally, and in the analytic process as a true summary of Belle's inner life. As I have indicated, Stoller makes it quite clear that the full text of the daydream only emerged over a period of years in the course of a long analysis; and he also indicates that the progress of the analysis, its success, can be calibrated accurately against the change in the daydream and its function in Belle's erotic life (the details of this process I will have to pass over). Hence we see the daydream emerge through the work of analysis and be slowly dissolved and transmuted by that work. It is therefore clear that this daydream could also be called a transference daydream, on analogy with Freud's concept of the transference neurosis: the structure of the relationship of the patient to the analyst created by the work of analysis, a structure into which all the symptomatic energies of the patient are channeled, only to be dissolved in the work of successful analysis. In this sense, as you have probably already guessed, I want to argue that the daydream is an *artifact* of analysis, in the same sense that the transference or transference neurosis is. Stoller is careful to give us the clues indicating that this is the case. He also has other reasons for letting this be seen.

The first of these reasons is that another, parallel, unsentimental education is recounted in this book—that of Stoller himself. Alongside the slow construction, reconstruction, and deconstruction of Belle's daydream, one finds the story of the development of Stoller's own theory of sexual excitement. In the introduction to the book—paginated in roman numerals to accentuate its set-off-ness from the main narrative—Stoller writes: "Some of my thoughts on the dynamics of excitement started years ago (before I shifted from an ordinary analytic practice to one primarily involved with gender disorders), especially with the analysis of the woman—Belle—on whom this book is focused. In the years since her analysis ended, my ideas have become clearer. They are tentative and need testing, especially by other analyses, but they are, I hope, a useful start" (xiii). In the course of the book, readers get more direct clues as to the development of Stoller's ideas in tandem with the analysis. But there is something else happening to Stoller that I must mention as well. He is, inevitably, being incorporated into the daydream, principally as the Director. Of course he is acutely aware of it; indeed, I could have told the story of the uncovering of Belle's daydream as completely consequent on his adoption of the position of, or the realization that he was, the Director in the daydream. Stoller writes a chapter entitled "Who Is Belle?"; I want to concentrate our attention on the question: 'Who is Stoller?' There are (at least) three Stollers:

> Stoller the Director in Belle's daydream;
> Stoller the Analyst of Belle's daydream; and
> Stoller the Theoretician of sexual excitement.

My hypothesis is that the emergence of Stoller as the Director—the emergence of the daydream qua transference neurosis—goes hand in hand with the emergence of Stoller's theory of sexual excitement. More ambitiously, I am suggesting that this sexual theory and the transference figure of the Director are internally linked, both genetically—in terms of their emergence over time—and conceptually. Stoller's theory of sexual excitement is internally linked to his taking on the transference role—of the Director—assigned him by Belle. In that sense, in so far as the transference is Belle's creative production, so is Stoller's psychoanalytic theory also her creative product.

Let us look at some evidence of this twin process. Stoller often casually refers to the fact that he is the Director: one can see that his role as the Director immeasurably facilitated the work of analysis, and he, like all good analysts, gratefully accepts the fantasy roles he is granted, while offering to Belle his interpretations of his role rather than acknowledging the gratifications she thinks him playing that role will accord her. Thus: "An element I missed for years . . . present but unknown in the transference until I caught it in the day-dream: the advantage in always being the victim was that her two Directors (he and I), though they mistreated her, concentrated wholly on her, unlike her undependable mother. . . . Before understanding this, I was simply her victim, puzzled as to why she was doing this to me. Insight is a relief" (79). At the end of the book, Stoller spells out how his coming to "know" what function he is performing is also the same as the completion of the analytic process of discovery: "Once we knew Belle's sexual fantasy, we could have predicted how she would deal with me in the analysis. . . . But until I knew the scripts, I could not understand what she was doing to me, and as long as one of her scripts was that I knew her scripts, she would not (consciously) understand what I did not know" (207). And he lists some of the things he was "doing" with her:

> She and I are performing a sexual drama, in which I—frozen and phallic—observe her as she humiliates herself by lying before me, saying what comes to her mind.

As does the Director, I sit behind her, sexually excited because I get turned on when humiliating women.

No analysis is actually in progress; there are only the two of us pretending—and both knowing it is simply pretense. . . .

As she lies there, she is a delectable morsel that I will devour with my eyes, if not my mouth. . . .

These scenarios . . . are, however, disguises for an underlying, essential, secret . . . script: no matter what happens, no matter how manifestly bad he is, the Director is steadfast. . . . Only late in the analysis did we discover that this fantasy, always present at some level, made her secretly comfortable with me. . . . All that counted, till we dealt with this belief, was that I not abandon her. (207–8)

Having shown how Stoller the Analyst manages his relationship with Stoller the Director, I now want to show how Stoller the Theoretician and Stoller the Director are linked. Throughout the book, small interventions, footnotes, or comments in parentheses reveal the intertwining of their functions. For example, to the sentence quoted above, Stoller adds a footnote: "Footnote: Perhaps the form all analysts take with their patients has in it a touch of Belle's Director" (269n5). In other words, he is suggesting that the structure of Belle's daydream and the analytic situation are inherently connected: to be an analyst is to play the part of the Director. This is a theoretical suggestion. There are others, more significant and more distinctively Stoller's:

The Director gradually became clearer, especially by the third and fourth years of the analysis. . . . During the two or more years in which this material emerged clearly, I began to think dimly some of the ideas now being made fully conscious in writing this book. I did not think them and then transmit them to the patient; it was the other way around in that, in time, she revealed the structure of the daydream and, with her associations, its roots. Meanwhile, I was showing her how she had transferred her daydream into our relationship, so that she believed she was there to resist being humiliated in my presence by excitement that was the result of my directorial sadism. (Erotic excitement felt for the analyst is not easy to analyze. It carries into the treatment a sense of reality, for it is actually experienced, not just a memory from another time, or an intellectualization. Or even an insight. Being a sensation, it is a truth of the body. It also, of course, fit into Belle's key fantasy, for my analyzing rather than sleeping with her was the equivalent for her of the Director's refusal to touch her, his exciting her from a distance without involving himself.) (76–77)

Stoller indicates, somewhat bashfully, the parallel process of becoming the Director and of developing his theories of sexual excitation. Other hints are less crafted. Here is one example; as I mentioned, the slow transformation of Belle's daydream constituted an important index of her overall cure. Stoller writes:

> A change in a variant script when, as the analysis progressed, her hostility was decreasing: a dirty-old-man is watching her masturbate. Instead of being humiliated, she has an assignment: to teach him (mother, father, stepfather, me*) about her excitement. If she teaches him clearly, he will know how to handle her erotic needs skillfully. She is on her way to becoming a child, a better role than that of victim. In the next phase, although the men may be depicted doing things she considers perverse, they do so to help, not humiliate, her. (86)

The slow evolution of the daydream provides a clear index of the changes induced in Belle's inner life: the Director or the dirty old man is being taught about her, so that his actual actions, which may not change very much, nonetheless are intended for her enjoyment rather than for her humiliation. This educative dimension is obviously one closely linked to the analysis Belle is undergoing, and to Stoller's role in it. So it is very striking to find the following footnote (as opposed to an endnote) attached to the reference to "me" (the analyst): "*My interest in analyzing her excitement was always in the analysis, years before I imagined writing on this subject" (86).

Why did Stoller add this note? On the face of it, he is reassuring the reader that Belle's new aim of teaching the observer does not derive from her knowledge that her analyst is interested in the theory of sexual excitement; he is disclaiming a prior interest in order to reassure the reader that *her* intentions are not contaminated by *his* prior interests. On the face of it, Stoller is telling the reader that *he had no particular interest in sexual excitement at the time of the analysis*, and that his interest in the topic was only kindled years after. He portrays himself as the innocent analyst, devoid of excessive curiosity or desire for knowledge, with *no prior theoretical commitments*. On the face of it, he is informing the reader that he had no interest in her sexual excitement *outside* of the analysis; only *in* the analysis, as *part* of the analysis, did such an interest arise. Of course, if he were to say that, even in the analysis, his interest in her sexual excitement was not aroused, he would be denying his own analytic function. The Director *must* pay attention, *must* be involved in Belle's sexual excitement for the scene to work. Thus in order to interpret the transference, Stoller had to become aware of and interested in her sexual excitement—to resist doing so would be a countertransferential resistance. If a patient deploys an erotic daydream in analysis, dreams it, replays it with the analyst as a protagonist, the analyst is professionally *required* to get interested. What Stoller is emphasizing is that he had no interest *other than* this strictly analytic interest inherent in *her* analysis.

So we are brought up short by wondering if this moment, when Belle's daydream began to shift from performing for the male onlookers to educating the male onlooker, including her analyst, is not *also* the moment when Stoller's extra-analytic interest in sexual excitement was born. Instead of just watching, as if he were the Director at the movies, he is now being educated through life, being required to play a part in Belle's education of the analyst. Remember how I was careful to point out Stoller's wide and flexible range of methods of working with people: as analyst, but also as professor, as psychiatrist, as ethnographer, as sympathetic collaborator in s-m groups, as interviewer of porn stars and directors. In other words, is this book the product of Stoller's education by Belle? Did Stoller add on to this analytic function that of being educated by Belle about sexual excitement?

Now we begin to see the epistemological point: Whose ideas are these anyway? I have argued that Belle's daydream, which seems so distinctively hers, is in fact a construction of analysis—a transference phenomenon. Now I am implying that this may also be the case for Stoller's theories of sexual excitement. Critics of psychoanalysis often put in question either the patient as a reliable source of data or the reliability of the analyst as a scientific observer. Sometimes these criticisms take the extreme and therefore clear form of accusing the analyst of suggesting *all* the data to the patient, or they take the form of seeing the analyst as the dupe of the patient—if you believe *that*, you'll believe anything! What I am suggesting is that the epistemological problem recognizes the partial truth in both of these criticisms, but takes the argument one step further. The transference, it should be remembered, was, in Freud's view, both real and not real. As Glen Gabbard put it recently: "Freud also stressed that transference love should nonetheless be treated as though it is unreal, a dictum that may appear confusing. . . . Neither analyst nor patient is likely to believe that the feelings are 'unreal.'"[9] The transference neurosis should be fully recognized as a genuine case of mutual irreality. And here we see the consequences: Belle formulates her own daydream so that it coalesces in a very productive way around Stoller the Director, while Stoller, through adopting the role of the Director, is able to develop a theory of sexual excitement. They have created a relationship in which both participate equally—if not in the same way—in the creation of something new.

We now come to the *Nachtrag* I referred to, with its epistemological bomb. Writing in the late 1970s in California, Stoller was acutely aware of the issues of confidentiality, privacy, and informed consent involved in medical treatment— the doctrine of informed consent, I can remind you, was introduced into law by a Kansas judge in 1961. Freud had solved these problems in his own way by pseudonymity, suppression or distortion of details, or a judgment that the subject of the case was indifferent to its publication; Stoller found a new solution to the problem of publishing a revealing case history of a patient.[10] On the surface it is a solution to a medico-ethical problem; what it contains, however, is of enormous consequence for the whole notion of psychoanalytic truth and its relation to transference and countertransference. The following is a long quote, but I think it is worth it—and it gives essential evidence for my argument:

To be sure that Belle's anonymity was preserved, I contacted her while writing this book and told her it would not be published without her complete approval. To do this, I asked if she would review every word of every draft. She has. Doing that became an addendum to the analysis, a new piece of work that focused on her responses to reading about Belle. In this way, I not only hoped that she would catch details that might otherwise reveal her identity but also that I would be ensured accuracy in all other respects. This process became for both of us a creative and surprising experience in which we could, from this new perspective, check each of our impressions, first of what had happened between us in the analytic process and second of what had happened in the unfolding of Belle's life from infancy on.

Not only were the fires of transference rekindled, but the book's reality in itself re-created a central theme of the erotic daydream. As you recall, the Director makes Belle reveal her excitement, as a display for an audience. And that, most literally, is what I do with this book. So not only was reality conspiring with fantasy, but she was in the dilemma of deciding what were her motivations in cooperating with me: to what extent would she give me permission, from the detached view that it was of use in the study of erotics, to tell her story, and to what extent would she do so simply because she was still under the daydream's spell. For surely it is unethical to get her permission if she is unduly influenced by a persisting, powerful transference effect. (It has been Belle's absolute right, till the moment the book is actually published, to stop the production simply by asking that that be done.) Only when we both felt this was worked through, when we knew it was depleted of erotic charge, were we at ease that I could proceed. I recommend this technique—of doing a piece of analysis with one's patient during the writing—to colleagues in a similar situation.

For Belle this experience was a valuable "review," in which, with full impact, she relived the analysis and was able to reconfirm the insights and changes that resulted. For me, it was, at the least, a test of reliability, in the sense that "reliability" is used by the experimental psychologist. Did the data re-emerge pretty much the same? Yes. Were there many or important corrections in the facts or perspectives? No. Nonetheless, it was an amusing and sobering experience for me to see where interpretations I *knew* were correct—they came out of me with that sense of spontaneous, enthused creativity that is the most fundamental confirming experience one can have—had been wrong. Wrong either in emphasis or in significance if not just flat-out wrong. (Of course, now I know, with the same enthused conviction, that *this time* my impressions are correct.) Not all of my truths, we discovered, were Belle's truths.

This experience I also recommend to colleagues. (217–18)

Among the many remarkable things in this passage is the criterion for publication of this book: Stoller thinks it unethical to publish if Belle's motive for agreeing is the result of the transference; the index of the disappearance of the transference motive is the depletion of the erotic charge—from what? From the new version of the erotic daydream, in which Belle's humiliation is on display not only to the Director, not only to her analyst, but to any reader of the book. At this point, I recall Freud remarking to Sándor Ferenczi that one could never eliminate unconscious structures—"we can bring about nothing more than exchanges, displacements; never renunciation, giving up, the resolution"—the most one could do was to change the balance of investments of those structures.[11] The shift from a small child consumed by murderous sibling rivalry to that older child's enthusiasm for dissecting live spiders and insects to her, on becoming an adult, practicing as a skilled surgeon is simply a change in the investments of the structures, not their elimination. In a sense, it is we readers who have the ethical obligation as well—to adopt the position of the director and be steadfast for Belle.

Looking at it from the point of view of the analyst, we see that Belle and Stoller collaborated to ensure that the publication of the book was entirely depleted of erotic charge—for her. We might say that the elimination of the erotic charge *completes* the analyst's transferential education, rather than *destroying* it. In this as in many things, Stoller hands over authority to his patient's version of the world—that version of the world that he had immediately spotlighted in the opening sentence: how she felt she was when the analysis began. By the end of his piece of clinical writing, we know how Belle felt she was when the analysis ended. And that transformation involved turning the analyst into someone who knew about sexual excitement. Otherwise the analysis would not have proven a transformation. Here we see something like an epistemological obligation for Stoller to write about Belle and to write about sexual excitement. If he had not published this book, her analysis would have been a failure.

In a sense, I am pointing out that Stoller's analysis of Belle provides a clear example of what has come to be called "countertransference enactments," defined as "a joint creation that depends partly on what the patient coerces in the analyst and partly on what is present in the analyst to respond to that coercion."[12] But the coercion here is mutual, so it is a transference-countertransference enactment. Stoller and Belle become acutely conscious of the sexual reciprocity of their relationship, its enactment of sexual intercourse: "I recalled that at that moment, it was as if she had offered an opening into which I could move; so I did. The metaphor became the feeling that I literally was allowed to move into her body, and yet her resistance was still present. But now it was resistance-plus-opening" (109). That is a fine description of sexual intercourse (including the anal overtones crucial to Stoller and Belle's enactment—remember the enemas); and it is a fine description of the discursive relationship between Stoller and Belle. But the more profound enactment is not that of the details of Belle's humiliation by the dirty old man in view of the Director, but the final bottom line that I have already quoted: "All that counted, till we dealt with this belief, was that I not abandon her." In her erotic daydream, she was safe because, in the end, she was the person watching the Director watch her being humiliated: "behind the scenes, a part of herself permits the excitement because it (she) knows that she, who is masturbating in the real world, is not literally the same as 'she' who is the suffering woman in the story. In the story, she is humiliated; in reality, she is safe" (68). Adopting the position of watching the watcher, thus having the watcher unaware of being watched, is ensuring that the Director not abandon her. Stoller, in writing this case history, enacted this fantasy for her. Not only did he not abandon her but he asked her to go over every word of every one of his sentences—she truly became the watcher watching the watcher. As Stoller remarks: "By the end of analysis, she knew (most of the time; like the rest of us, she had her episodes of backsliding) that whatever may have been done to her in the past, in the present she is the victim of herself: she can be free if she knows that she is the Director's director" (87). The book that "they" produced—and in this "they" we can include Belle the Watcher, the Director's Director, Belle the Humiliated Woman, the Director, Stoller the Analyst, and Stoller the Theorist—is the final monument to and proof of Stoller not having abandoned her. He as the author, she as the "case" are there, in print, for all time to come. Safe.

In conclusion, I want to rerun my argument very quickly in a different area of Stoller's case. To do so, let me return to a term that became an important part of Stoller's theoretical vocabulary, and which we have come across already: *scripts.* "A script or scenario," Stoller writes, "is a story line—a plot—complete with roles assigned to characters and a stream of action. When a script is conscious, it is, if private, either a spontaneous, unwilled emergence or a daydream" (xiv–xv). Belle's daydream is in this sense a master script. And she lived her life in such a way that everybody participated in her scripts all the time—until she was woken from this lifelong dream by her analysis. But it is not only Stoller who thinks in scripts; it is quite clear that Belle's language is also that of the script: is not the Director, the key figure of her erotic life, someone with a script, someone who directs a script, someone who follows a script?

It is at this point that I want to interject something not present in Stoller's text. Gossip being gossip, someone—and I can't remember who—told me who Belle's mother actually was in real life. I am not going to reveal that, and not only because there's a strong possibility that the information was false—even though I am a firm believer in the virtues and epistemological function of gossip. But I will say that she was a Hollywood film star of the 1930s, 1940s, and 1950s. This much is almost evident in Stoller's description of the histrionic mother, surrounded by admirers, always in connection with famous men, going off for long periods for work. But this piece of information makes us wonder some more about Belle's inner life. When she constructs the daydream of the Director, it is modeled in part on her stepfather, her mother's second husband who, as the husband to a film actress, has very much the domestic role of a Director. Read as part of the culture of Hollywood, the book takes on a different resonance entirely. In particular, one part of the relationship between Stoller and Belle comes into sharp profile: the fact that much of his countertransferential work was done in the field of the visual. By that I mean that he found he had to pay enormous attention to her visual performance for him: in particular, the clothes she wore and the exhibition of her body that she then performed using those clothes. He referred to these in various ways: as "storms of exhibitionism" (131), a "skirt-waving attack" (134); "the big show" (144), which he thought was the single aspect of her that was "so loony that it struck me as little short of delusional" (144): "Sitting up on the couch, she would pose herself in pin-up-girl fashion, leaning against the adjoining bookcase with back arched, legs drawn up in knee-chest position, skirt fallen so that her thighs and part of her buttocks showed (but with the genital area always 'casually' covered by clothing and posture)" (144). Stoller's observations of Belle's clothes and how she used them, how she would wiggle her skirt up her thigh, then move it down again, began to be drawn into the analysis; almost against his will, he found that he *needed* to comment on her clothes, on how she was lying on the couch, on what point up her thigh her dress had reached, and so on. He became the analytic voyeur: "She came in one day dressed deliciously. . . . She kneeled on the couch in order to go on display. . . . She was so edible that I wondered to her if the couch was not functioning as a table on which she was serving herself up. The suggestion gave her an acute erotic attack, for which she was ripe" (154). I quote these passages not only because they are unconventional analytic observations, as Stoller knew better than anyone, but because they give a sense of the dramatic and filmic ambience in which the analysis was conducted. This analysis was about femininity, and it was about Belle's conception of the feminine as the given to be seen; but it was also conducted with a mother who was a film actress, in a town in which the given to be seen is not only a culture but also an economy. This analysis was not only conducted in Hollywood—it was also Made in Hollywood. And out of that collaboration between Stoller, Belle, and the ambient culture came the theory of scripts, closely linked to Stoller's critique of the theory of repetition that focused on the medium of representation by insisting that analytic repetition was not true repetition but simulation.[13] For me, the confirming evidence of Stoller being obliged to collaborate at all these levels—transference-countertransference, theory, and cultural ambience—comes in his very final comment on the success of Belle's analysis: "At the end of her analysis she was still likely to begin the process of communicating important messages to me by being a character in a staged scene, with me as another player. Only after I caught on and described to her the plot and our parts in it was she able to understand an interpretation" (217). Each patient and analyst jointly construct a common language in which to conduct their relationship—a private vocabulary, a set of references, images, and signifiers, so that a pause, a chuckle, a grunt, and a couple of words come to speak volumes. Certainly Stoller never persuaded Belle to speak the language of classical psychoanalysis—a language that he, too, grew to detest—being open to the "ludic drift" of language and reference to which the analyst should surrender himself.[14] But in this passage we see that, right to the very end, Stoller was obliged to speak Belle's language—the language of drama, of characters, of plots, and of enactment. Stoller took some of these terms and turned them into theory. Or to put it another way, in putting forward his theory of scripts and sexual excitement, he is still speaking her language—he remains true to the inner content and means of her representation of her daydream. In this sense, he did not so much transmit psychoanalytic knowledge *about* Belle as *infect* readers with the terms and frame of reference of Belle's analysis. We should not forget—though we can if we wish—that they were "originally" Belle's terms, from her daydream, from her way of being in the world, and maybe even from her mother and her world. In not forgetting, we reenact the ethical obligation of the analyst in taking up the epistemological functions assigned to the Director.

What, as a result of reading Stoller's case history of Belle, can we add to my preliminary dogmatic remarks concerning the function of case histories in psychoanalysis? Can we credit Stoller with the achievement of objectivity, or is his case one more instance of the analytic self-exposure that gives epistemic pleasure—the Director's pleasure—but no more? Stoller strived for the variety of objectivity achieved by intersubjective agreement—a rare enough event, psychoanalytically, but still subject to the suspicion that it derived from the logic of the folie à deux. Given the materials on which this essay is based, how could it provide *more* than that? And is it not more plausible that my argument cannot achieve even this, and will instead simply repeat the elements of the staging of the scene of Belle's desire and of Stoller's (analytic) desire, simply redeploying their transference-countertransference relationship in a displaced mode, even conniving with and exacerbating the exhibitionism and voyeurism so conspicuous in their work? If Stoller can muse that "perhaps the form all analysts take with their patients has in it a touch of Belle's Director" (269n5), we may readily acknowledge that the same will apply to the readers of psychoanalytic case histories.

This theoretical and interpretative aporia may not be soluble as it stands. Perhaps we should turn to the crude but crucial criterion of cure and progress. Stoller had mapped the progress in the development of Belle's daydream and regarded it as a key indicator of the progress of the analysis. As we have seen, Belle becoming aware of her powers as the "Director's director" was for him the sign of this progress. But an alternative formulation was: "The Director, however, never returned" (87). We can thus view it in one of two ways: the director had been fired, got the sack, because a higher authority stepped in (the Director's director). Belle had her way with the Director, and then even had her way—as final arbiter—with Stoller. Getting rid of the Director is a neat allegory for the end of the analysis too. And if Belle was satisfied, if Stoller got what he wanted (the case, the theory, the satisfaction of a cure), who are we to know better? It is always open to the reader, my reader, to judge otherwise. And another conclusion is also a feasible option: that this psychoanalytic case, like others, through its failure to secure more than a parade of its own remarkable dynamics, demonstrates how the dispatch of the Director can also be seen as the exit of the analyst.

NOTES

An early version of this essay was given as a paper to the Centre for Psychoanalytic Studies, University of Essex, June 5, 1996; I would like to thank Karl Figlio for the invitation and for his comments. A later version was given as a public lecture when I was resident as a visiting research scholar at the Getty Research Institute for the History of Art and the Humanities, Los Angeles, on May 11, 1998; I would like to thank Page du Bois in particular for incisive comments on that occasion and for penetrating reflections on that version of the essay by Amanda Pustilnik. A yet later version was given to a seminar in the Program in the History of Science, Princeton University, on March 26, 2001; I would like to thank Liz Lunbeck for her comments on that occasion and on earlier versions (particularly in an e-mail of January 11, 1999), and the late Gerry Geison for his response. Finally, I would like to thank Sybil Stoller for an absorbing and informative discussion on May 23, 1998 concerning her late husband's work.

1. Sigmund Freud, "Analysis Terminable and Interminable," 1936, in *The Standard Edition of the Complete Psychological Works of Sigmund Freud*, ed. James Strachey, 24 vols. (London: Hogarth, 1953–74), 23:248.

2. See Robert J. Stoller, "A Contribution to the Study of Gender Identity," *International Journal of Psycho-Analysis* 45 (1964): 220–26.

3. Robert J. Stoller and Ira S. Levine, *Coming Attractions: The Making of an X-Rated Video* (New Haven: Yale University Press, 1993), 4.

4. Ibid., 3.

5. Ibid., 5.

6. Robert J. Stoller, "Naturalistic Observation in Psychoanalysis: Search Is Not Research," in *Presentations of Gender* (New Haven: Yale University Press, 1985), 2.

7. Robert J. Stoller, *Sexual Excitement: Dynamics of Erotic Life* (1979; London: Maresfield Library, 1986), xvi. Future references to this work will be made parenthetically in the text.

8. The internal quote by D. Shakow is taken from A. Wolfson and H. Sampson, "A Comparison of Process Notes and Tape Recordings: Implications for Therapy Research," *Archives of General Psychiatry* 33 (1976): 559.

9. Glen O. Gabbard, "Sexual Excitement and Countertransference Love in the Analyst," *Journal of the American Psychoanalytic Association* 42 (1994): 1086.

10. Other solutions have since been found. Sybil Stoller informed me that her husband had sometimes made recorded tapes of sessions, which his secretary would

type up for him to write from. When he died, the University of California, Los Angeles, destroyed these tapes in order to protect patient confidentiality.

11. S. Freud, "Letter from Sigmund Freud to Sándor Ferenczi, January 10, 1910," in E. Brabart, E. Falzeder, and P. Giampieri-Deutsch, *The Correspondence of Sigmund Freud and Sándor Ferenczi, Volume 1, 1908–1914* (Cambridge, Mass.: Harvard University Press, 1994), 123.

12. Gabbard, "Sexual Excitement," 1084; see T. J. Jacobs, "On Countertransference Enactments," *Journal of the American Psychoanalytic Association* 34 (1986): 289–307.

13. The genealogy of the concept of script is clearest in the line developed by the transactional analysis of Eric Berne in his *Transactional Analysis in Psychotherapy: A Systematic Individual and Social Psychiatry* (New York: Grove, 1961) and in his *Games People Play: The Psychology of Human Relationships* (New York: Grove, 1964). Berne first introduced the notion in a reasonably casual fashion in one of his few conventional analytic papers, "The Psychological Structure of Space with Some Remarks on *Robinson Crusoe*," *Psychoanalytic Quarterly* 25 (1956): 557, alongside another notion, that of "unconscious protocol." Another important analyst at work in the 1970s who made ample use of the notion of script, particularly for sexual perversions, was Joyce McDougall. See, for example, her "Primal Scene and Sexual Perversion," *International Journal of Psycho-Analysis* 53 (1972): 371–84. It may be of relevance that I date Belle's analysis to the 1960s, probably the early 1960s, well before the notion of script had become widespread outside of Carmel, California, where Eric Berne practiced. When Stoller introduces his concept of "scripts, scenarios, scenes, daydreams, and fantasies" (xiv)—whose connotations he recognizes overlap—he cites not Berne but J. Laplanche and J.-B. Pontalis, *The Language of Psycho-Analysis* (1967; New York: Norton, 1974), 318: "Even where they can be summed up in a single sentence, phantasies are still scripts (*scénarios*) of organised scenes which are capable of dramatisation— usually in a visual form. The subject is invariably present in these scenes; even in the case of the 'primal scene,' from which it might appear that he was excluded, he does in fact have a part to play not only as an observer but also as a participant, when he interrupts the parents' coitus."

14. A phrase I owe to a personal e-mail communication from Page du Bois, May 15, 1998.

"To Exist Is to Have Confidence in One's Way of Being": Rituals as Model Systems

CLIFFORD GEERTZ

Though rarely remarked, and never, so far as I am aware, explicitly discussed, the association of the names of famous anthropologists with the names of particular rituals they, more or less accidentally, happened to have studied (and which have become, thereby, equally famous) is both very common and, once established, apparently permanent. Bronislaw Malinowski and the Trobriand *kula*; E. E. Evans-Prichard and the Zande poison oracle; Victor Turner and the Ndembu *mukanda*; Gregory Bateson and the Iatmul *naven*; Franz Boas and Ruth Benedict and Kwakiutl potlatches; Roy Rappaport and Maring pig feasts; Max Gluckman and Swazi mock rebellions; Géza Róheim and Australian subincisions; James Mooney and Plains Indians Ghost Dances; Sir James Frazer and the priestly immolation at Nemi; Mary Douglas and the Lele pangolin; Marshall Sahlins and the Hawaiian deification of Captain Cook: to each, it seems, his or her defining rite.[1] As someone who sees himself fated to be remembered as the author of a brief piece on the Balinese cockfight, I am concerned to know why this identification of ethnographers with the rites and ceremonies they describe, analyze, and, perforce, publicize should be so strong and so persistent. Why does it exist? What does it signify?

In part, of course, it signifies the importance of ritual, and especially of collective, obligatory ritual in the sort of societies—"traditional," "primitive," "simple," "savage," "tribal," "archaic" (all antiquated, presuming terms, never really descriptive in the first place)—anthropologists have, at least until recently, mostly studied. Since the days of John McLennan's, William Robertson Smith's, and James Frazer's elaborate theorizations about "the merry sacrificial feast of totemism" toward the close of the nineteenth century, ritual, together with kinship (in totemism, of course, the two are fused, along with our third obsession, the incest taboo, into a single all-encompassing phenomenon), has been the very index of comparative ethnology—its natural subject and defining kind.[2]

Indeed, "in the beginning"—that is, from the 1870s and 1880s onward, when Frazer, Smith, and the rest began to work, and thereby launched the modern conception of anthropological analysis—ritual, understood as a universal phenomenon, an intrinsic, pervading feature of the "life of men (and women) in groups," was not just a means through which to grasp the morals and mentalities of exotic Others but itself constituted the sum and substance of those morals and mentalities. In the passage rites, the calendrical ceremonies, the communal feasts, and the burnt (and unburnt) offerings of "simpler peoples," as well as in their dances, exorcisms, ordeals, games, and mutilations, the so-called ritual school of classicists and anthropologists—Jane Harrison, Francis Cornford, A. M. Hocart, Émile Durkheim, latterly Lord Raglan, and René Girard—saw the foundations, "real" and "historical," of myth, obligation, social order, cognition, authority, and faith alike. Ritual was less a model of something than the something as such, the *fons et origo*, and the essence as well, of the human form of life. *Homo ritualis.*

The vicissitudes of this "in the beginning was the deed, and the deed was culture" point of view have often been traced, and its limitations, severe to the point of falsity, canvassed. We need not attend again to them here.[3] The decline of so-called originist explanations in anthropology, evolutionary, diffusionist, degenerative, or otherwise, and their replacement by functionalist, structuralist, and hermeneutical explanations, that is, systemic ones, shifted the role of ritual in anthropology as a marking topic. From being an object *to* study, a sort of sociological natural kind, it became an object to study *with*, a means and a modality: the microscope, not the bug under it.

This is obliquely put, and perhaps unnecessarily enigmatical. But what it means is that rather than a theory of ritual as such, a theory that was at the same time a theory about the general, universal character of something called "man," or "humanity," or "human nature," anthropologists have been more concerned, since the 1930s or so, with what particular, specifically ordered rituals can tell us about particular, specifically shaped societies, and about the sorts of lives such societies support. Like the obese mouse or the deformed fly of contemporary molecular genetics, armshell exchange, pig sacrifice, pilgrimage, shamanic healing, and, yes, the cockfight, have, in someone or other's molding hands, become what looks rather like what historians of science call a model system: "An object or process selected for intensive research as an exemplar of a widely observed feature of life."[4]

The correspondence is not exact. Rites are not organisms, or even molecules; certainly, they are not genes. And anthropology is, despite the efforts of evolutionary psychologists, memeoticians, and other Darwinian fundamentalists to make it so, not biology. But a number of the characteristics ascribed by biological researches to model systems—specificity, typicality, materiality, complexity, accessibility, standardization, boundedness, individuation, narratability, and "self-reinforcingness" ("the more a model is studied, and the more perspectives from which it is understood, the more it becomes a model system") seem ascribable, with rather little twisting and pushing about, to ritual.[5] Or, rather, they seem applicable to the use of ritual by anthropologists to construct what passes for knowledge in our rather catch-as-catch-can, case-in-point field.

If ritual, at least for most anthropologists now, is one or another sort of model system, rather than a unitary phenomenon in need of a sovereign explanation, what sort of thing, then, is it, or more exactly, particular instances of it, supposed to model? The answer would seem to be just about anything any particular anthropologist is concerned, at this time or that, to explicate or explain. Social structure, worldview, psychological development, artistic expression, economic exchange, cognitive style, gender definition, technology, cultural stability, cultural disruption, representational thought, personal identity, political authority, and, of course, the shapes of the divine and the duties thought to flow from them have all been viewed through the lens—magnifying, reducing, or distorting—of one or another ritual or ritual complex taken to be particularly well-suited for the purpose.

In any given case, a number of such matters may, of course, be in play at once; but the intellectual center of gravity is readily enough discerned in even the most complex and sophisticated of such studies. Malinowski's concern, not to say obsession, with the ordering of practical, everyday life in the face of uncertainty, Turner's with the workings of transformative emotional experience in the reenergizing of social life, Evans-Prichard's with the nature of causal reasoning, or Bateson's with the forms and processes of communicative interaction all appear most clearly in their analyses of ritual. The work of these people, figures themselves exemplary, is fairly generally known, only too often summarized. But at the risk of reviewing familiar material, let me take that of the two of them, Malinowski and Turner, generally considered to be paradigm instances of, respectively, the functionalist and the structuralist approaches to ritual, and describe their work, or rather redescribe it, in terms of the model-system idea, "the model-system model." I want to reframe it in a "thinking-in-cases," "exemplary-narrative" form so as to have some referent instances when I turn, further on, to my own, much alike but rather different, hermeneutical approach to things.

It is not unusual for an anthropologist's statements of theory and method to do more to obscure his or her ethnographical work than to illuminate it; but Malinowski abused the privilege. His so-called, self-called functionalism represented cultural institutions as instruments in the service of biological and psychobiological necessities (nutrition, reproduction, the regulation of emo-tion, the focusing of desire), while his work, and especially his work on ritual, was trained on, as he put it, "the victory of tradition and culture over the mere negative response of thwarted instinct"—a moral concern, not a mechanical one.[6]

The central importance of ritual, which he essentially identified with magic and saw as a sort of occult technology ("a body of purely practical acts, hedged round by observances, mysteries, and taboos, performed as a means to an end") in the moral life of, as he put it, "man," runs through the whole of his work on the Trobriands—two volumes of fussy agricultural rites in *Coral Gardens and Their Magic*, elaborate funeral ceremonies in *Baloma*, mass renewal feasts in *Myth in Primitive Psychology*.[7] But it is in his most famous work on the kula, that enormous circle of reciprocal gift giving "lying on the borderland between the commercial and the ceremonial . . . expressing a complex and interesting attitude of mind," that the specificities of his approach are most forcibly ex-pressed.[8]

In the kula, the continuous movement of ceremonial objects—necklaces clockwise, armshells counterclockwise—around an enormous, interisland, intertribal circle of ritualized exchange, a long sequence of precisely staged magical rites, during canoe building, during sailings, navigatings, and landings, during the exchanges themselves, pace the system, give it form, and, energizing its participants, keep it going:

> [In the *kula*] the consecutive progress of work and of magic are insepa-rable just because, according to native ideas, work needs magic, and magic has only meaning as an indispensable ingredient of work.
>
> Both work and magic are directed toward the same aim; to construct a swift and stable canoe; to obtain a good Kula yield; to insure safety from drowning, and so on. [Magic] consists in [particular bodies] of rites and spells associated with one enterprise, directed towards one aim, and pro-gressing in a consecutive series of performances that have to be carried out in their proper place. [This fact] is of the greatest theoretical impor-tance because it reveals the nature of the relation between magical and practical activities and shows how deeply the two are connected with one another.[9]

Magic—a stream of specifically focused offerings, purifications, omen readings, taboos, oaths, petitions to ancestors—models the moral content of practical life, the normative dimension of the everyday. "Primitives," as Malinowski still called them, are centrally concerned with the rational mastery of their surroundings. Rather than representing a world apart in which reality is dematerialized and overcome, magic proves an essential element in effecting such mastery—not because it suspends natural connections (even the "natives," as he also still calls them, weeding their yams and caulking their canoes, do not believe that), but because of its capacity to shape and sustain effort in the face of uncertainty, weakness, error, and natural calamity.

"What," Malinowski asks, "are the cultural functions of magic [again, read ritual]?

> [All] the instincts and emotions, all the practical activities, lead man to impasses where gaps in his knowledge and the limitations of his . . . power[s] of observation and reason betray him at a crucial moment. Human organism reacts to this in spontaneous outbursts, in which rudimentary modes of behavior and rudimentary beliefs in their efficiency are engendered. Magic fixes upon these beliefs and standardizes them into permanent traditional forms. [It] supplies . . . man with ready-made acts and beliefs, with a definite mental and practical technique, which serves to bridge over the dangerous gaps in every important pursuit. It enables man to carry out with confidence his important tasks, to maintain his poise and integrity in fits of anger . . . hate . . . despair . . . anxiety. The function of magic is to ritualize man's optimism, to enhance his faith in the victory of hope over fear. Magic expresses the greater value for man of confidence over doubt, of steadfastness over vacillation, of optimism over pessimism.

"I think we must see in it," he concludes, "the embodiment of the sublime folly of hope, which has yet been the best school of man's character."[10]

Malinowski, a nonbeliever, and something of a positivist, worked in a South Pacific island chain among navigators; Turner, a Roman Catholic, and something of a Marxist, worked on a Central African forest plateau among hunters.[11] Malinowski focused on exchange ceremonies; Turner focused on rites of passage. Malinowski celebrated the energies of practical life; Turner, the transformation of that life and those energies into forms of existence transcending the merely ordinary, mundane, accepted, and everyday. Naturally, they had rather different conceptions both of what ritual was—expressivist "magic" in the one case, transfigurative "drama" in the other—as well as what it was that ritual modeled: "The dogma that hope cannot fail nor desire deceive" in the Trobriands; "the generic and essential human bond . . . 'men in their wholeness, wholly attending'" in the Ndembu.[12]

There is no way to summarize, or even to order, the vast, and vastly detailed, ethnographic corpus—more than a thousand pages of it—that Turner produced on what he called "the Ndembu ritual process." But there runs through the whole of it a clear and unvarying, and, as it progresses, an increasingly explicit, frame of analysis. Departing from the work of the French anthropologist Arnold van Gennep on so-called rites of passage, Turner centered his attention on the form such rites characteristically take and the effect they characteristically have.[13] Passage ceremonies (marriage, initiation, burial, dispute settlement, first fruits) "indicate and constitute" transitions between what he calls social states (sickness-health, war-peace, childhood-adulthood, commonality-kingship, scarcity-plenty, life-death—the usual binaries of structuralist analysis).[14] Such rites are "a process, a becoming . . . a transformation."[15] He compares them to phase changes, water being heated to boiling point, or a pupa changing from grub to moth, and he sees them as setting the terms and possibilities of collective life. It is in the movement of people, or more exactly of groups of people, through the dense allegories of ritual, "the forest of symbols," that human community is formed, reformed, and held in place.

That movement, at once a dramatic, often enough melodramatic, public act and a shaking private experience, has both a shape and a substance, a visible dialectic. Following, again, van Gennep's lead, Turner sees passage rites as having, in the nature of the case, three distinct stages: separation, marginalization, and aggregation. The first phase comprises symbolic behavior

> signifying the detachment of the individual or group either from an earlier fixed point in the social structure, from a set of cultural conditions (a "state"), or from both. During the [second] intervening . . . period, the characteristics of the ritual subject are ambiguous; he passes through a cultural realm that has few or none of the attributes of the past or coming state. In the third phase (reaggregation or reincorporation) the passage is consummated. The ritual subject, individual or corporate, is a relatively stable state once more and . . . [again] has rights and obligations vis-à-vis others of a clearly defined and "structural" type; he is expected to behave in accordance with certain customary norms and ethical standards binding on incumbents of social position in a system of such positions. (Turner, *The Ritual Process*, 94–95)

It is the second, intervenient, stage of liminality (from Latin *limen*, "threshold"), and the out-of-category condition its disarrangements and dissolutions create, *communitas* (from Latin, *communitatus*, "diffuse, undifferentiated sociality"), that Turner takes to be the heart of "the ritual process."

"Liminal entities" (that is, persons and groups of persons in a liminal state— Turner also refers to them as "passengers" or "threshold people") are neither

here nor there, neither the one thing nor the other. "They are betwixt and between the positions assigned and arrayed by law, custom, convention, and ceremonial" (95). They are literally outside structure; indeed, they are in an amor-
phous, ambiguous realm of "anti-structure." Their situation is likened, in the
symbolic dramas into which they are, more or less forcibly, plunged, "to death,
to being in the womb, to invisibility, to darkness, to bisexuality, to the wilderness, [or] to an eclipse of the sun or moon" (95). They may be dressed as mon-
sters, as paradoxical, unclassifiable beings. They may go naked to demonstrate
that they have no status, no property, nothing to distinguish them from others
in the same between-state state. They are secluded, humbled, punished, rendered passive. They are covered with filth, forced to lie motionless, surrounded
by masked mummers threatening and mistreating them. "It is as though,"
Turner says, "they are being reduced or ground down to a uniform condition to
be fashioned anew" (95).

It is this ground-down, raw condition of humanity that Turner calls communitas:

> What is interesting about liminal phenomena . . . is the blend they offer
> of lowliness and sacredness, of homogeneity and comradeship. We are
> presented, in such rites, with a "moment in and out of time," and in and
> out of secular social structure, which reveals, however fleetingly, some
> recognition (in symbol if not always in language) of a generalized social
> bond that has ceased to be and has simultaneously yet to be fragmented
> into a multiplicity of structural ties. These are the ties organized in terms
> . . . of caste, class . . . rank hierarchies . . . segmentary oppositions. . . . It
> is as though there are here two major "models" for human interrelated-
> ness, juxtaposed and alternating. The first is of society as a structured,
> differentiated, and often hierarchical system of politico-legal-economic
> positions . . . separating men in terms of "more" or "less." The second,
> which emerges . . . in the liminal period is of society as an unstructured
> . . . undifferentiated . . . community, [a] communion of equal individuals
> who submit together to the general authority of the ritual elders. . . . It is
> . . . a matter of giving recognition to an essential and generic human bond,
> without which there could be *no* society. (96–97)

Given this worm-to-butterfly conception of the passage from abandoned structure through antistructure to emerging structure, a passage fundamentally exis-
tential in character, Turner sees it just about everywhere ritual is prominent: in
millenarian movements, in coronations, in mendicant orders, in Shakespeare's
The Tempest, in Bengali devotionalism, and, of course, in its underlying para-

digm the Eucharist. He invokes everyone from Martin Buber and Leo Tolstoy to William Blake and Anna Freud in his support. Even the hippies and the Hell's Angels get a mention. But, as with Malinowski, the main proving ground for Turner's theory is an enormously detailed description of a personally observed tribal institution of extraordinary reach and complexity: mukanda, the Ndembu circumcision of adolescent boys.

There is no need, and no possibility, of detailing Turner's description of the mukanda here. It goes on for over a hundred pages, dense with medicines, symbols, beliefs, spells, settings, offices, paraphernalia, all particularized in terms of specific persons immersed in specific events. At its core a typical ceremony in which adolescents are "reborn as men after a symbolic death," it involves dozens of what Turner calls "episodes," stretching over many months and involving, in one role or another, virtually the entire Ndembu population.[16]

The circumcision itself—a grand, ecstatical drama of drums, blood, fear, pain, and wailing coming early on in the sequence—is but part of an extended process designed to remove boys from the mollycoddle realm of the mother-centered domestic family and make of them fully masculine beings, capable of being hunters, lovers, husbands, fathers, fighters, and chiefs. The symbolic forest through which the mukanda winds, Turner says, is not (as standard structuralism would have it) just a set of *pensée sauvage* cognitive classifications for ordering the Ndembu universe. It is also, and more critically, "a set of evocative devices for rousing, channeling, and domesticating powerful emotions . . . hate, fear, affection . . . grief" (42–43).

If kula magic is an engine of action, mukanda melodrama is a school for passion: "The whole person, not just the Ndembu 'mind,' is existentially involved in the life and death issues with which [it] is concerned" (43). Like psychotherapy and street gangs, the mukanda makes something happen.

But is all this enfoldment of social and psychological life in the symbolic gesturings of ritual but a feature of so-called tribal, backward, or disappearing societies, such as the Trobriands and the Ndembu apparently are? Is not "modern" society, if not wholly lacking in the likes of canoe magic and rebirth imagery, at least fairly well free of their domination—essentially secular, rational, morally transparent? Are not the scraps of serious ritual still about in churches and cults but atavisms, relics of a vanished past hanging on in the present out of cultural inertia and an uneven transition to a disenchanted conception of reality? Is the notion of ritual-as-a-model-system without application in today's secularized, size-up-and-solve world?

Put in so preemptive a way, the answer seems dictated. One is naturally inclined to respond with an irritated and resisting "no!" The we-they othering contrast seems extreme and old-fashioned, itself a relic of departed binaries— *Gemeinschaft-Gesellschaft*,[17] savage-civilized, mechanical-organic, backward-advanced, traditional-modern. The idea that so-called tribal societies are precursory to our own, "the world we have lost," is not much in fashion anymore. "We've got the goods, but they've got the morale," no longer seems to sum up the situation.

Yet for all that, it is clear that elaborate, inclusive ritual systems pacing the whole or the greater part of life and directing its course are hardly prevalent in the contemporary world (though caste may be one example, sharia another). The forest of symbols has been everywhere thinned or cut down, and it is a good deal more difficult to see what use there might be in looking into the sort of dispersed and peripheral ritual—from Halloween and birthday parties to the Superbowl and loyalty oaths—that does exist. What are we to make of our own conjurations and passages?

In addressing this question, I want to begin at what might seem a surprising place: with some recent work, mostly by psychologists, but by anthropologists and neurologists as well, on what the culture theorist Louis Sass has called "madness and modernism."[18] In such work, the proposition that both Malinowski and Turner advanced (and, in their several ways, most other contemporary anthropologists who have worked on ritual have advanced as well)—that the sense that life has meaning, significance, point, and import, that "it adds up," is, more often than not, fragile, uncertain, and under pressure—is extended to the ambiguities and intricacies of modern life. Confidence in the depth and substantiality, the "reality," of one's world, and of one's way of living in it, and thus of one's self, is, for humans anyway, not a natural given, but a social, cultural, and psychological achievement, recurrently threatened, occasionally destroyed. It is that achievement, and that threat, which, among us as among the Trobriands and the Ndembu, ritual can be seen to model.

Sass, a clinical psychologist by trade, working mainly with schizophrenics, begins by comparing various characteristics of the work, that is the sensibilities, of archetypal modern and postmodern artists and writers—Franz Kafka, Giorgio de Chirico, Samuel Beckett, Antonin Artaud, Alfred Jarry, James Joyce, Stéphane Mallarmé, Salvador Dali, Guillaume Apollinaire, Pablo Picasso, André Breton, Arthur Rimbaud, Charles Baudelaire, Virgina Woolf, Robert Musil, Djuna Barnes, Ludwig Wittgenstein—with the acts and statements, the "symptoms," of his patients. In both, there is projected a suffusing sense of lostness and alienation—hyperreflexivity, doubt, dehumanization, desocialization, disconnection, derealization, a paralysis of the will to act, a disorienting, an impoverishing, "unworlding of the world":

> [Far] from indicating a lowering or shutting-down of conscious aware-
> ness, many . . . experiences in schizophrenia actually involve forms of
> "hyperreflexitivity" and alienation . . . accompanied by a diminishment
> of what the phenomenological psychiatrist Eugene Minkowski . . . called
> "vital impulse." This condition is characterized by an exaggeration . . . of
> self-consciousness . . . an acute awareness of, and disengagement from the
> grounding frameworks and assumption that normally serve as the taken-
> for-granted foundation of organized action and expression and of coher-
> ent modes of experience.
>
> . . .
>
> These frameworks and assumptions are cultural in origin . . . and their
> smooth functioning is a precondition of normal social existence. [These]
> core features of schizophrenia . . . show remarkable resemblances to some
> of the key aspects of modern culture and society—which is itself charac-
> terized by "wholesale reflexivity" and by associated forms of detachment
> and alienation from commonsense reality.
>
> The schizophrenic is an anomalous yet at the same time exemplary
> figure: a person who fails to adopt social practices or internalize the cul-
> tural frameworks that are essential to normal social life, yet whose failure
> to do so often tends to illuminate these very frameworks and processes of
> internalization, which at the same time typifying some of the most dis-
> tinctive features of the modern age.[19]

The resemblance of this "modern age state of things," mental and cultural, to the gaps and impasses that threaten Malinowski's outrigger navigators with doubt and lack of confidence, a disabling collapse of nerve, or to the antistructural, in-and-out-of-time removal from the ordinary and the everyday that Turner's liminal entities experience in their trips through communitas, is perhaps ap-parent. The "unworlding of the world" is, it seems, not confined to the inven-tive, the mad, or the disenchanted.

We get in this way a rather different, more general, less culture-specific con-ception of what ritual-as-a-model-system models, what it is a prototype, case, or example *of*. It is not just practical action, category change, and the like, but modes of being-in-the-world, forms of life, that, properly studied, these cul-tural Drosophila throw into relief.

Sass cites Wolfgang Blankenburg, a German psychiatrist, to the effect that "the fundamental trouble" afflicting the schizophrenic is, in the words of one of Blankenburg's patients, "the loss of natural self-evidence," the loss (now Sass) "of the usual commonsense orientation to reality, with its unquestioned sense of obviousness and its unproblematic background quality which allows a person to take for granted . . . the elements and dimensions of [the] shared world."[20] This patient, a young woman, describes herself as without "the evidence of feelings," as lacking as "stable position or point of view" on life and how one ought to live it. She says everything strikes her as novel, "strange," "peculiar." She feels "outside," "beside," or "detached," as if "I was regarding from somewhere outside the whole movement of the world." "What am I missing, really?" she asks— "something small, [odd], something important, but without which one cannot live. . . . To exist is to have confidence in one's way of being. . . . I need support in the most simple everyday matters. . . . It is . . . natural evidence which I lack."[21]

Schizophrenia is, doubtless, in good part organically rooted, and cultural devices such as magic or passage rites are unlikely to prove very effective in overcoming it and the pain that it causes. (Though they may help sustain those afflicted with it to get on with their fractured existence.)[22] But lack of confidence in one's way of being, "the loss of natural self-evidence," "the diminishment of the vital impulse," the pervasive sense that one is "missing something small, odd . . . important . . . without which one cannot live," is a general potential in human experience. Whether it be brought on by weakness, threat, or isolation, by moving from adolescence to adulthood or by burying the dead, by the disabling confusions of modern life, or by some genetically programmed neural misfire in the brain, the capacity to make sense of one's self, of one's world, and of one's self in one's world is everywhere and for everybody somewhat under pressure.

Ritual is not the only way in which this pressure is, when it is, countered, either in Germany or for the Trobriands. But it is a good place to begin, in either case, if one wants to capture the effort to have confidence in one's way of being "in all [its] complex uniqueness while at the same time rendering [it] in a generically analysable form."[23] And it is for that reason that so many anthropologists, anxious to break into other imaginaries, have found in one or another instance of it—a circumcision, a cockfight, Jonesville, Waco, or Heaven's Gate, the national communitas of Anzac or Bastille Day, Princess Diana's funeral or the World Trade Center towers commemorations, the practical magic of psychotherapy—a strategic and convenient object of attention, a model system of a particular way of engaging the real, of worlding the world.

NOTES

This is a revised and extended version of a paper originally presented to a history of science workshop titled "Model Systems, Cases, and Exemplary Narratives" at Princeton University on November 9, 1999, organized by Norton Wise, Angela Creager, and Elizabeth Lunbeck. I am grateful to them and to the workshop participants for their comments and criticisms. A version was published in Italian as "Esistere è avere fiducia nel proprio modo d'essere. Rituali come sistimi modello," in L. Cimmino and A. Satambrogio, eds., *Antropologia e interpretazione. Il contributo di Clifford Geertz alle scienze sociali* (Perugia: Morlacchi Editore, 2004), 211–30.

1. Bronislaw Malinowski, *Argonauts of the Western Pacific: An Account of Native Enterprise and Adventure in the Archipelagoes of Melanesian New Guinea* (London: Routledge, 1922); E. E. Evans-Prichard, *Witchcraft, Oracles and Magic among the Azande* (Oxford: Clarendon, 1937); Victor Turner, *The Forest of Symbols: Aspects of Ndembu Ritual* (Ithaca: Cornell University Press, 1967); Gregory Bateson, *Naven: A Survey of the Problems Suggested by a Composite Picture of the Culture of a New Guinea Tribe Drawn from Three Points of View* (Cambridge: Cambridge University Press, 1936); Roy A. Rappaport, *Pigs for the Ancestors: Ritual in the Ecology of a New Guinea People* (New Haven: Yale University Press, 1968); Max Gluckman, *Rituals of Rebellion in South-East Africa* (Manchester: Manchester University Press, 1954); Géza Róheim, *Australian Totemism: A Psychoanalytic Study in Anthropology* (London: Allen and Unwin, 1925); James Mooney, *The Ghost-Dance Religion and the Sioux Outbreak of 1890* (Lincoln: Nebraska University Press, 1991 [1896]); James George Frazer, *The Golden Bough* (London: Macmillan, 1890); Mary Douglas, *The Lele of the Kasai* (London: Oxford University Press, 1963); Marshall Sahlins, *Historical Metaphors and Mythical Realities: Structure in the Early History of the Sandwich Islands Kingdome* (Ann Arbor: University of Michigan Press, 1981).

2. W. Robertson Smith, *Lectures on the Religion of the Semites*, 3d ed. (New York: Ktav, 1969). On "the merry feast," see George W. Stocking, *Victorian Anthropology* (New York: Free Press, 1987), 15.

3. For general reviews, see Catherine M. Bell, *Ritual: Perspectives and Dimensions* (New York: Oxford University Press, 1997); Catherine M. Bell, *Ritual Theory, Ritual Practice* (New York: Oxford University Press, 1992).

4. M. Norton Wise et al., "Charge to the Workshop on Model Systems, Cases, and Exemplary Narratives," Princeton History of Science Program, September 1999, unpublished manuscript.

5. Ibid.

6. Bronislaw Malinowski, *Magic, Science, and Religion and Other Essays* (Boston: Beacon, 1948), 5.

7. Ibid., 70, 87. Bronislaw Malinowski, *Coral Gardens and Their Magic* (London:

Allen and Unwin, 1935); Bronislaw Malinowski, *Myth in Primitive Psychology* (New York: Norton, 1926). In all of Malinowski's work the terms *ritual* and *magic* are essentially interchangeable.

8. Malinowski, *Argonauts*, 513.

9. Ibid., 414.

10. Malinowski, *Magic, Science, and Religion*, 90.

11. The Ndembu, like the Trobrianders, are subsistence cultivators, but hunting, like interisland exchange for the Trobrianders, is the focus of their cultural life, especially for males.

12. Victor W. Turner, *The Ritual Process: Structure and Anti-structure* (Chicago: Aldine, 1969), 97, 128. All future references to this work will occur parenthetically in the text. The internal quotation comes from the Catholic Mass, on which Turner's approach to ritual is modeled.

13. Arnold van Gennep, *Rites de passage* (Paris: E. Nourry, 1909).

14. Victor Turner, *The Forest of Symbols* (Ithaca, N.Y.: Cornell University Press, 1967), 93.

15. Ibid., 94.

16. Ibid., 152.

17. The German words mean "community" and "society," respectively, and were coined by Ferdinand Tönnies in *Gemeinschaft und Gesellschaft: Abhandlung des Communismus und des Socialismus als empirischer Culturformen* (Leipzig: Fues, 1887).

18. Louis A. Sass, *Madness and Modernism: Insanity in the Light of Modern Art, Literature, and Thought* (New York: Basic Books, 1992).

19. Louis A. Sass, "'Negative Symptoms,' Common Sense and Cultural Disembedding in the Modern Age," in *Schizophrenia, Culture, and Subjectivity: The Edge of Experience*, ed. Janis Hunter Jenkins and Robert John Barrett (Cambridge: Cambridge University Press, 2004), 303–28. It needs to be noted that Sass's approach is not reductionist: he does not identify symptoms and cultural expressions; he compares, connects, and reflexively redescribes them.

20. Wolfgang Blankenburg, *Der Verlust der Naturlichen Selbstverstaendlichkeit: Ein Beitrag zur Psychopathologie Symtomarmer Schizophrenia* (Stuttgart: Ferdinand Enke, 1971), as quoted in translation in Sass, "Negative Symptoms," 306; Sass, "Negative Symptoms," 305.

21. Blankenburg, *Der Verlust*, as quoted in translation in Sass, "Negative Symptoms," 306. With the exception of the word "odd," this is Sass's translation. Geertz translated *komisch* as "odd" whereas Sass chose to use "funny."

22. For rituals as "sustaining" (rather than "curing") chronically suffering individuals, see Vincent Crapanzano, *The Hamadsha: A Study in Moroccan Ethnopsychiatry* (Berkeley: University of California Press, 1973).

23. Wise et al., "Charge to the Workshop."

Democratic Athens as an Experimental System: History and the Project of Political Theory

JOSIAH OBER

WHAT IS A HISTORICAL CASE STUDY GOOD FOR?

Can the political history of classical Athens legitimately be regarded as a case study—an experimental system or exemplary narrative, useful for investigating various aspects of democracy and related phenomena? I hope to show that the answer to that question is yes, but first it seems necessary to pose an analytically prior question: What is the *goal* of the investigation? What precisely is the use-value of the case? What, in short, is the exemplary narrative supposed to be good for?

Those questions may make a proper historian a bit queasy. She or he might reasonably ask: Need history be good for anything other than historical knowledge itself? The point is that as soon as I say that the historical experience of, for example, classical Athens is an experimental system, I seem to have abandoned the historian's tried and true ground of supposing that it is sufficient to say that Athenian history is worth studying for its own sake. I admit to a personal fondness for this sort of traditional historian's propriety. So, as a preliminary move, let me plant a stake in the ground (in Greek terms, a *horos*) by saying that I actually do suppose that Athenian history is interesting for its own sake. I will try to advance the argument for the use-value of Athens as an exemplary narrative as far as possible beyond that marker. I may be forced to give some of the ground in front of the stake by the end of the day. But the stake of "Athens interesting for its own sake" is the point behind which I would not retreat without making a desperate stand.[1]

The most obvious answer to the "exemplary for what?" question is, I suppose, "exemplary for other periods of history." Which is to say that the case of democratic Athens might usefully be deployed for comparative methods of argument in historiography because, via close attention to a range of similarities and differences, the history of Athenian political development might tell us something meaningful about other periods of human experimentation with democracy and related forms of political organization. I think that such a claim, carefully framed, could be supported. I am very much in favor of comparative arguments in history, and I have quite often made reference to exemplary narratives drawn from the history of early modern Europe, for example, in order to make a point about antiquity—especially in seeking to extend the range of interpretive possibilities that might be considered by historians of antiquity.[2] But I do not have anything new to say about the method of comparative history—the advantages and problems with the comparative approach have been a matter of discussion for a long time among historians (students of antiquity will be likely to think, for example, of Plutarch's *Parallel Lives*), and I can see no very obvious way that the theory of "the case" would allow me to add anything fresh or original to those discussions.[3]

Is it reasonable, then, to look *outside* the discipline of history in seeking to establish a use-value for the Athenian case? On the face of it, yes. The use-value of a given biological model system is not de facto limited to a particular academic discipline. The mouse is used in experiments by both medical pathologists and behavioral psychologists. *Drosophila* have proved just as useful for evolutionary biologists investigating the geographic diffusion of visible traits as they have for molecular biologists who analyze the structure of individual genes. Arguably, the conceptual category *model system* connotes interdisciplinary use-value. So if the exemplary historical narrative qua *experimental* system is indeed analogous to a biological *model* system, its use-value should potentially exceed the disciplinary domain of history per se. I will return, at the end of this essay, to similarities and differences between model and experimental systems.

I propose, then, to ignore comparativism within the discipline of history, and to argue instead that historical narratives can prove usefully exemplary, can serve as experimental systems, outside the discipline of history as such. The most obvious extrahistorical field for which the case of Athenian democracy might seem to have use-value is political sociology. Classical Athens offers an example of a fairly rare and yet analytically significant sociopolitical phenomenon: direct participatory democracy as the primary system of governance for a society that was fairly large (i.e., too big to be explained in terms of face-to-face interactions), quite complex (literate, socially differentiated), and politically independent (i.e., not subject to superior external or internal political authority).[4] There has been at least sporadic interest among political scientists in classical Athens as a case study in direct democracy. Some of my own work on Athens has been aimed at convincing political scientists that the Athenian case challenges the universality of the so-called iron law of oligarchy, which asserts that participatory democracy will inevitably degenerate into oligarchy for structural reasons involving leadership roles.[5] But I propose leaving political sociology to one side for the time being in order to focus on the possibility that the historical case of democratic Athens may be usefully employed within a different academic field: ethical and political theory. I will argue that the Athenian case is useful for normative thinking about problems of justice, rights, and obligations.[6]

I am quite aware that asserting that a particular historical case could usefully inform normative theory means confronting a familiar obstacle—and the obstacle may be regarded by some students of normative theory as simply impassible. The brick wall in question is defined by David Hume's famous assertion that "one cannot derive an 'ought' from an 'is.'"[7] And so, presumably (perhaps a fortiori), not from a "was." An expansive interpretation of the scope of Hume's dictum might, therefore, seem to write history out of ethical and political philosophy.

Before exploring ways to get around the wall, we should ask whether there is any very good reason to *want* to get around it. After all, the experimental system must, by definition, model something that relevant disciplinary practitioners actually care to learn more about. If consideration of the historical case does not offer something needed (or valued) by normative theory, it seems senseless to proceed with the endeavor of inviting political theorists to take cognizance of the case of classical Athens. It may come to seem akin to urging astrophysicists to begin experimenting with fruit flies, on the grounds that experiments on *Drosophila* have materially advanced other scientific fields.[8]

Claiming that experimenting with historical cases could usefully advance the project of normative theory is not, on the face of it, much easier to sell than fruit flies to physicists. Political theorists are very interested in texts written in the historical past (by, for example, Thomas Hobbes, John Locke, and John Stuart Mill), but they tend to regard past historical practice as irrelevant to their scholarly project, and they have long tended to regard "historicism" as a gross methodological error.[9] Theorists use the charge of historicism somewhat loosely and in ways very reminiscent of the ways some historians use the term *essentialism*: In each discipline-specific argumentative context, she or he who fails to defend her- or himself against the damaging charge must regard the argument offered as effectively refuted. This casual and pervasive use by political theorists and historians of terminological shorthand to produce knockdown methodological refutations has, I think, unnecessarily limited intellectual intercourse between political theorists and historians.[10] By way of personal anecdote: When I first came to Princeton in 1990, I frequently attended meetings of the Political Philosophy Colloquium (which centered, then as now, on the general problem of liberalism and its critics) and meetings of the Davis History Seminar, then focused on imperialism and colonialism. I was graciously welcomed in both forums, but I quickly found that framing interventions in each venue, without leaving myself open to knockout punches of historicism, on the one hand, or essentialism, on the other, proved a severe challenge—even for someone professionally involved, as I was and am, in the study of rhetoric.

Now, the strong Humean imagined above might say that historicism is not just a scare term like *essentialism*, that it serves a function more serious than simply shoring up current canons of disciplinary taste. The Humean might suggest that historicism for the political theorist is equivalent to the historian's concern with anachronism, that is, a shorthand for a methodological error so basic as to render any claim infected by the error invalid a priori. For instance, in response to a historian's charge of essentialism, I might respond: "Yes, indeed, my argument does assume that some particular and modest aspects of human behavior are innate, rather than socially constructed and contingent, but I hope to convince you that making this sort of essentialist assumption will allow me to explain a fuller range of the historical phenomena we are seeking to understand." I suppose that most historians would be willing to hear me out, judging the potential validity of my "essentialist" assumption on the basis of its explanatory power or lack thereof. On the other hand, I would expect to be regarded as simply ridiculous by all serious historians if I claimed that "my argument does in fact assume that persons living in an earlier age intuitively anticipated the meaning of much later developments and shaped their behavior accordingly." Or, a fortiori, that "my argument does in fact assume that earlier event A was caused by later event B." My point here is that I believe historicism (as I am using the term) in political theory is actually more like essentialism in history and less like anachronism, that is to say, an acknowledged danger and an interpretive tendency easily taken too far, but not a vicious methodological error.

The dominant (within the Anglo-American philosophical tradition) approaches to problems of justice and ethics may be broken out (somewhat artificially and schematically) as Millsian consequentialism/utilitarianism, Kantian deontology, Hobbesian social contract theory, and Habermasian deliberation.[11] And each of these approaches does indeed respect Hume's dictum in that each avoids basing moral claims on assumptions that are in any strong sense historical. In each case, however, concerns arising from fairly recent historical experience and (at least in the case of Mills and Hobbes) from the study of the history of Athenian democracy played a role in the original formulation of the particular approach to the problems of justice and ethics.[12] That is to say, in each of the examples mentioned, case history played some role in the formulation of the moral problem to be addressed, although not in the formal solution of the problem (the establishment of the normative claim itself, grounded in each case in ahistorical premises).[13] My point here is that one may accept the force of Hume's dictum and still allow history to play a role in framing the agenda of problems that normative theory seeks to address.

So history would seem to have a familiar (if delimited) place in moral philosophy. Yet I think it is fair to say that the bulk of recent work within each of the powerful traditions of moral philosophy noted above remains quite studiously ahistorical. Many, if not most, modern moral and political philosophers (at least those working within the dominant Anglo-American analytic tradition) remain quite safe from attack by those wielding the methodological scare term *historicism* in that historical cases are kept marginal in the arguments they advance.[14]

Contemporary moral philosophers tend to advance and to defend their substantive arguments by means of "hypothetical cases" or "thought experiments" rather than historical cases—and these hypotheticals and thought experiments typically tend toward either the bland or the bizarre.[15] The purpose of the hypothetical is to reduce the issue under discussion to its simplest, most mundane or extreme, and thus most accessible elements. This allows for the proposition in question to be tested by reference to the moral intuitions of the audience. For example, the proposition "one must never break a promise" might be tested by the following hypothetical case: John has promised Mary that he will go to a movie with her tonight. But meanwhile, John's family is endangered by a flood. Should John take Mary to the movie, or save his family from the advancing flood waters? Since the obvious intuitive answer is "save the family," the original proposition seems to require revision. Of course not all hypotheticals need be quite so banal or obvious. John Rawls's well-known thought experiment involving choice making by individuals ignorant of their actual life circumstances and personal preferences (which are hidden behind a "veil of ignorance") is an example of just how powerful and productive an apparently bizarre thought experiment can be when it is well constructed and worked through in detail. There has, of course, been a great deal of important work on moral and political philosophy done by such methods in the past generation. Yet I will polemically assert that the perpetuation of a systematic methodological insulation of political theorizing from the messiness of history leads to a potentially serious problem.

The problem is that the ahistoricism of the hypothetical and the thought experiment threatens to deracinate normative claims from the more complex aspects of lived experience. The emphasis on bland or bizarre hypotheticals, rather than on genuinely complicated historical circumstances, means that normative claims may have relatively little purchase on the complicated practical issues that confront many people, much of the time, and in many aspects of their lives.[16] Thus to the extent to which the political theoretical project succeeds in the rigorous exclusion of history, it also threatens to distance itself from the hope of making practical ethics a genuinely meaningful category.

So, to put the issue in bald terms: because much of the common experience of ethical and political life is overtly embedded in history, denying the relevance of history works against any hope of arriving at a robust understanding of ethical behavior or "the political." Taking this claim on board, which means adding the historical case study to the hypothetical and the thought experiment in the arsenal of tools regularly employed by theorists when exploring political and ethical problems, does indeed entail the risk that there will be some contamination of normative claims by concerns that are frankly historical (whether or not they are historicist) in nature. Scrupulous Humeans may feel that the risk is not worth taking. But for those who suppose that the practical benefits of returning political philosophy to a closer dialogue with history outweigh the risk of blurring the ought-is distinction, the question becomes something like this: given that an engagement with history must be part of any serious investigation of justice, rights, and obligation, what sort of historical case study will best allow us to test and perhaps to extend the scope of our normative intuitions?[17]

First, by analogy with the biological model system, we should obviously seek a historical case that will allow us to explore those particular normative issues that we actually care about. This means that we would expect the case to manifest those features that we suppose must be part of any account of justice, for example, that will make sense to us, here and now. And so I would propose that at a minimum, the case should have some bearing on the values of liberty, equality, dignity of the person, and reciprocity.[18]

Next, like a biological model system, we would want the historical case to be generally accessible to practitioners. This means it should be sufficiently rich and fully articulated to genuinely flesh out our intuitions. Yet is should not, at any given point in our investigation, be so overwhelmingly complex as to require a great deal of highly specialized research in order to decipher the general outline. The case must be "true to life," but that liveliness must not require that political theorists become specialist historians in order to appreciate it.

Finally, unlike the biological model system, I think that we would want the historical case to be *near* enough to us that we could actually identify empathetically with the historical agents in question, as well as remaining *distant* enough for us to gain some critical distance from those agents. It should allow us to learn from both difference and from similarity.[19]

Does the historical Athenian experience with democracy fit this bill of particulars? I should emphasize at this point that what I am primarily referring to is the actual practice of Athenian democracy and the public rhetoric typical of Athenian democratic institutions, rather than Athenian political theorizing as such. It is not, therefore, the familiar question of "did Plato have some ideas about justice that a modern theorist might regard as valid or at least as interesting?" But rather, "does the history of democracy in classical Athens yield a case that would allow modern political theorists to test their current assumptions about the relationship between values and practices?"

In various ways, democratic Athens does seem to offer the sort of experimental system that political theorists who want to test normative intuitions against an exemplary narrative case might find useful, for (at least) the following five reasons:

(1) Classical Athens manifested a distinctive form (direct democracy within a complex, literate, midsized society) of a general type of human organization (democratic governance). Both specific form and general type are indeed of great interest to many contemporary theorists. Both have obvious contemporary real-world relevance. Although we will need to be very careful about equating ancient to modern systems of value, it is certainly reasonable to assert that the values of freedom, equality, dignity of the person, and reciprocity were central to the way Athenian democrats (and Athenian critics of democracy) thought about their political lives.

(2) Democratic Athens has a finite history, with a fairly clear beginning (ca. 508/7–462 B.C.), a well-documented middle (ca. 462-322 B.C.), and an end (although there is less agreement about this last, democracy is certainly gone after the Roman general Sulla sacked the city of Athens in 88 B.C.).

(3) There is a large, quite detailed, but finite and thus manageable, body of primary evidence for understanding the case. Most of the primary evidence for Athenian democracy is widely available as standard text editions and standard epigraphical corpora readily available on the Internet. Therefore testing new theories is relatively easy. Theories which fail to account for relevant available evidence can be falsified and thus decisively rejected. Yet this potential for falsification still leaves room for competing explanations for specific historical phenomena since there are a variety of useful and meaningful ways of organizing the existing evidence.[20]

(4) Despite its great chronological distance from us, Athens and its history does (historically speaking, at least) matter to us. Athens has been used by political writers as a case study of the relationship between political development and ethical norms since at least the middle decades of the fifth century B.C., and that relationship has been particularly important as a topos of Western political

thinking since the late Renaissance.[21] The Athenian case was especially to the fore in the seventeenth through nineteenth centuries, when most of the major accounts of moral philosophy discussed above were first developed.

(5) There is currently quite a high degree of agreement among specialists in the history of democratic Athens about basics: the chronology, structure, and function of primary institutions, the role of public rhetoric and political ideology. There is still considerable room for debate, however, among proponents of explanatory theories that are variously centered on the institution, the individual agent, or on discourse and ideology.[22]

USING THE ATHENIAN CASE

Having claimed that employing historical case studies may prove methodologically useful for political theorists, and that the Athenian case offers a potential example of an exemplary narrative useful for political theorizing, it is only fair to point out that for the past fifteen years or so, I have been working within the still ill-defined interdisciplinary terrain between classical Athenian history (institutional, social, political, intellectual) and political theory. Throughout, I have simultaneously been attempting to refine the case itself (i.e., trying to clarify the history of Athenian democracy, explaining how and why it worked in practice) and running political theoretical "experiments" on the case in an attempt to better understand various political and ethical problems of contemporary relevance.[23] The experiments seek to show how various phenomena familiar in one way or another to us from contemporary political and ethical discourse (e.g., rights) change when they are resituated within a system based on a strong conception of citizenship and a political commitment to participatory democratic decision making.[24]

It is because of this recursive relationship between the case and the theory that Athens should be regarded as an experimental system rather than a model system: I am, as it were, struggling to refine my strain of mouse while simultaneously running experiments on it. This means that the model is not really a fixed quantity; its external properties are always to some degree in flux, and so I cannot reduce my attention to a single variable: my understanding of the historical mouse changes somewhat as a result of each theoretical experiment performed on it.

It would admittedly be methodologically tidier to stabilize the model system (i.e., finalize the exemplary narrative) and then do the experiments (i.e., test normative conceptions against that fixed narrative). And something like that approach is, I think, taken by some political theorists who have looked to Athenian democracy as an example (positive or negative) of a type of politics, and who have adopted for the purpose a rather thin and wooden textbook account of Athenian history. The problem with this "model system/fixed exemplary narrative" approach is that the history does not end up doing much real work. It remains insufficiently thick (rich, fully articulated) to allow a genuine testing of the starting-point normative intuition. And so history serves as little more than an illustration or a straw figure for demonstrating the inherent validity of the original intuition.[25]

My own approach assumes that the case itself (the narrative of Athenian history) will inevitably tend to get richer and more fully articulated as a result of the experiments run on it (the testing of intuitions). I suppose that a virtue may arise from this necessity in that (if we remain very sensitive to what is going on) the recursive process can lead to a genuine and substantive gain in our understanding of history *and* will extend the range of our normative theorizing. The process remains interactive and pragmatic: there is no "final" state for either our understanding of the case itself or for our normative thinking about politics and ethics. Yet at various points in the ongoing process, we should be able to stand back from it and give a reasoned account of the current state of our understanding of both history and theory.[26]

It would certainly be foolish to claim that Athens can usefully model all or even most of the phenomena that a political theorist might be interested in exploring. There are some very real limits to the range of political phenomena that can effectively be analyzed through the Athenian case because there are some modern values, institutions, and practices that have no very meaningful analogues in Athenian practice. For example, the system of party politics, as we understand it on the basis of the historical experience of the Anglo-American world (and elsewhere) since the eighteenth century, has no real analogue in Athens. The attempt to identify structures resembling political parties in democratic Athens will eventuate either in a definition of "party politics" that is so expansive and baggy as to be analytically meaningless, or in a serious misreading of Athenian political practice. This point was made clearly enough by historians of Ancient Greece from the 1960s through the early 1980s: their work invalidated earlier attempts to explain the operations of supposed Athenian political parties. They showed that the analytic category "Athenian party politics" simply did not effectively organize the very considerable evidence for individual and group behavior in Athenian political life: too many phenomena apparent in the sources had to be left out of the party politics story, and too much behavior not mentioned by the source evidence had to be imaginatively restored.[27]

If party politics is the negative example drawn from the field of political science, there are, I believe, a good number of positive examples of political-theory experiments that can be run on the Athenian case. I list four here, drawn mostly from my own work.

(1) The relationship between the political regime and its critics is an important one in contemporary political theory. The salutary role of criticism is obvious enough when the regime in question is overtly vicious and the critics are calling for liberation from tyranny. But the legitimate role of critics is more difficult to explain when, for example, critics of a constitutional democracy call for its revolutionary overthrow. One possible approach is to divide critics into "good internal" critics (those who call on the constitutional regime to be true to its own highest principles) and "bad external" critics who reject the values embraced and nurtured by constitutional democracy.[28] A consideration of the Athenian case usefully complicates the basic scenario by introducing the problem of demotic discursive hegemony and the need for the critic of democracy to create a discursive space, a new vocabulary of politics, in order to gain a voice with which to carry out his critical project. Creation by critics of a discursive space can then be seen as an essential part in the maintenance of a robust conception and practice of free and frank speech—and as playing an important role in preventing hegemonic democratic political language from closing off the potential for innovative responses to evolving political challenges.[29]

(2) The concept of rights stands at the center of modern liberal political thought. But what exactly is the ontological status of a right? Where do "natural" or "inherent" rights come from, and how do they come to be "inalienably" attached to individuals? It has long been accepted that there is no exact ancient counterpart for the modern concept of right, although recent work by Fred Miller on Aristotle's *Politics* asserts that something quite similar to a conception of a natural right was implicit in the Aristotelian language of political entitlement.[30] A consideration of the Athenian case points to certain legal immunities (especially immunity from outrage: *hubris*) as important elements of Athenian law. These immunities have strikingly "rightslike" attributes in that they protect the immunity-holder against threats to his or her freedom, equal standing, and dignity. Notably, in Athenian law, immunity from outrage was (in principle) legally extended to *all* residents of Athenian territory (free and enslaved, male and female, adult and child), not just to citizens. But the preservation of the widely distributed immunity right was not via recourse to state authority mustered through a constitutional guarantee, but via the active exercise of participation rights restricted to the enfranchised citizenry (i.e., adult native males).[31]

(3) The meaning of citizenship is an increasingly important topic in contemporary political theory, and one of the key issues involved with citizenship

is naturalization: through what processes does one become an enfranchised member of the political community? An exploration of the public rhetoric associated with naturalization processes in Athens demonstrates the conceptual centrality of reciprocity. Some modern commentators have been struck by the overtness of the exchange relationship: Athenian grants of citizenship were accompanied by a frank expectation of ongoing public benefactions. The Athenian citizenry (*demos*) clearly believed it to be imperative that newly enfranchised citizens acknowledge that naturalization was a jealously guarded privilege, one that could only be granted by the *demos* through special decree in public assembly. The point was to ensure an appropriately high level of reciprocal gratitude on the part of the recipients, who would be morally required to pay back their collective benefactors with reciprocal benefactions. The rhetoric of naturalization thus reveals the extent to which the citizen body as a whole was integrated through a sense of reciprocal obligations. This recognition has much to say to theorists concerned with the prevalence of public virtues, with how citizen virtue is manifested.

(4) The inculcation of public virtue, via formal educational institutions, is a related concern of contemporary theorists. Some recent theorists have worried about the relationship between teaching the sorts of civic virtues regarded as essential for the maintenance of a democratic state and the protection of the rights of minorities (especially religious minorities) to educate their children as they wish.[32] Turning to the Athenian case, one is immediately struck by the *absence* of formal educational institutions, an absence critically remarked on by Athenian philosophers. But the Athenians clearly did suppose that the polis provided a thoroughly adequate education in public virtues and citizenly decorum—through the procedural practices of democratic institutions and the substantive decisions of public bodies. Decision-making institutions—especially the courts, assembly, and council—were regarded as fulfilling especially important educational roles, but public religious rituals, military service, and the everyday life of the public square were also regarded as playing an important part in educating the Athenians.[33]

In each example cited above, a familiar issue within contemporary political theory emerges as a recognizable issue within the Athenian case. But in each example, the familiar framing of the contemporary issue must be reconfigured as it is run through the mill of Athenian political history. The democratic Athenians are seen to share certain of the contemporary theorist's concerns, but to have addressed those concerns in ways that are likely to be unfamiliar and perhaps even disconcerting to contemporary assumptions. The point is not that modern liberals embrace freedom, equality, dignity, and reciprocity and Athenians failed to do so, but that in each case Athenian conceptions of the value in question had very different roots, and their loyalty to the value was differently manifested in practice. Throughout, the Athenians proved more frankly pragmatic in their approach to political problems, and less concerned with establishing unassailable foundations for belief and behavior than are most modern political philosophers.

But of course, not only were Athenians more overtly pragmatic in their democratic commitments than most moderns but they were also much less concerned with the universalization of core values, either within Athenian society or beyond the borders of Athenian territory. Full citizenship was always limited to adult males, and ordinarily limited to natives. Citizens held a monopoly on most participation rights and enjoyed privileged access to other social goods. Chattel slavery was prevalent—although scholars continue to debate just how prevalent it was, and how important slave labor was in the larger Athenian economy. Systematic moral concern about the institution of slavery seems to have been rare. Most Athenian women were not cloistered within their homes (as was once widely assumed), but their freedom of movement was considerably less than that of Athenian men. Women lacked most participation rights, and in most respects they lacked equal (or any very meaningful) standing before the law. Athenians tended to be frankly chauvinistic: they often regarded other Greeks as dull-witted, and non-Greeks as lacking the capacity for creating self-governing communities.

One of the major unresolved issues within Athenian sociopolitical history remains the extent to which the privileges of the citizens were established and maintained through the systematic oppression of noncitizens. Was the edifice of Athenian democracy—with its quasi-modern concern with free speech, political criticism, the extension of legal immunities to the weak against affronts offered by the powerful, and so on—constructed on a strictly hierarchical denial of political privileges to noncitizens? Or was oppression simply part of the cultural background? Whereas some classical scholars have suggested that democracy in Athens necessitated an *increase* in the oppression of slaves and women, others have suggested that the historical tendency of the democratic system was to blur the lines of established social distinction by extending a growing set of "citizen-like" privileges to at least some categories of noncitizens.[34]

This debate will be resolved (if ever) by Greek historians working on the detailed evidence for Athenian social relations. But questions about the relationship between democracy and the wider Athenian society are being debated (at least in part) because Greek historians are aware of the value of the case to contemporary political thought. We continue to care as much as we do about Athens in part because Athens is, and long has been, an exemplary historical narrative for participants in Western civilization.

CONCLUSION: SIMILARITY AND DIFFERENCE

In conclusion, I return to the question with which I began: What is the nature of the understanding we expect to gain from the establishment of a historical case study? When we "experiment" on the case, do we hope for knowledge that is essentially true, or can we be satisfied with contingent truth? Are we seeking values that are transcendental or culturally situational? Are we looking for regularities across cases, or are we seeking to explain what is distinctive about the particular case?

I suppose that the very act of studying the usefulness of exemplary historical narratives for nonhistorical disciplines signals a dissatisfaction with the epistemological dichotomies enacted by the language of the set of questions laid out in the previous paragraph. History, on some important level, is about difference: to study a past society is to engage with contingency, distinction, and imperfection. Normative theory, by contrast, is committed to ideals and universality, with establishing what is just, right, and good everywhere, always, and for everyone. Yet in neither case can this be the whole story. When history looks only to difference, it risks being reduced to a banal list of discrete and unrelated facts: this happened, then that, then some other thing—ad infinitum. Presumably it is the hope of finding some way to transcend this regress into meaningless distinction that leads historians to employ analytic models and methodological paradigms derived from other disciplines.[35] Meanwhile, moral theorists confront the concern that actually completing the normative project of "getting it right" would result in claustrophobic homogenization. The loss of the diversity of experience and capacity for surprise and change seems, at least to some, a terrible price to pay for moral certainty.[36]

Although I have focused here on political philosophy, the difference-similarity point can, I think, be generalized: it simply makes no sense for practitioners of nomological disciplines (i.e., those concerned more with establishing regularities than with celebrating contingency and diversity) to turn to the sort of rich and articulated exemplary historical narratives I am advocating here if their only goal is more securely established "laws"—whether these be laws of economics, social behavior, governmental organization, international relations, or whatever. The complex narratives that I suppose are potentially valuable as historical case studies are unlikely to yield a simple corroboration of basic laws—not without a prior stripping away of the complex features that make the case interesting to work with in the first place.

The only really useful historical case is one rich and fully articulated enough to challenge the capacity of any general law to adequately model the fullness of human experience. Nonhistorians who do engage with thick exemplary historical narratives do so, I suppose, because they sense that when framing general laws, it is important to be forcefully reminded of just how messy human experience really is. If nothing else, the historical case can serve to preserve within our conceptual schemata some salutary clutter, a healthy residue of surprise and puzzlement, a modesty about the capacity of our intuitions and behavioral laws to capture the peculiar reality of our actual lives.

NOTES

1. The military metaphor is deliberate and meant to recall the Greek association of boundary markers, analogies from warfare, and the concept of taking a political stand. See Josiah Ober, "Greek Horoi," in *Methods in the Mediterranean: Historical and Archaeological Views on Texts and Archaeology*, ed. David B. Small (Leiden, the Netherlands: Brill, 1995), 91–123. I have tended to cite my own work in this essay, for reasons that I seek to explain below.

2. See, for example, Michael Zuckerman, "The Social Context of Democracy in Massachusetts," *William and Mary Quarterly* 25 (1968): 523–44, an influential article explaining the peculiarly democratic political practices typical of villages in colonial Massachusetts by reference to the impossibility of maintaining order without the voluntary cooperation of adult males whose property fell below the statutory voting limits. Or consider the relationship between the refusal of the National Assembly to obey the French king's order to disperse and the subsequent storming of the Bastille; see Josiah Ober, *The Athenian Revolution: Essays on Ancient Greek Democracy and Political Theory* (Princeton: Princeton University Press, 1996), chap. 5.

3. Of course just because I do not myself see how the theory of the case would allow a rethinking of comparative approaches to history does not mean that I assume that such a rethinking is impossible.

4. The other classic case of direct democracy, the Swiss *Landsgemeinde*, suffers from the analytic defect of being ultimately subject to federal authority. See Mogens H. Hansen, *The Athenian Ecclesia: A Collection of Articles, 1976-1983* (Copenhagen: Museum Tusculanum Press, 1983), 207–26.

5. On the iron law of oligarchy, see Robert Michels, *Political Parties: A Sociological Study of the Oligarchical Tendencies of Modern Democracy* (New York: Collier, 1962); and critique by M. I. Finley, *Democracy Ancient and Modern*, 2d ed. (London: Hogarth, 1985). See also my own work on this: Josiah Ober, *Mass and Elite*

in Democratic Athens: Rhetoric, Ideology, and the Power of the People (Princeton: Princeton University Press, 1989); and Ober, *Athenian Revolution*, chap. 3.

6. Although self-described political theorists tend to have their appointments in political science departments, and self-described political philosophers in philosophy departments, and there may be some difference in the language in which "theorists" and "philosophers" typically frame their arguments, I intend no analytic distinction in this article between political theory and political philosophy.

7. David Hume, *A Treatise of Human Nature* (Oxford: Oxford University Press, 1978 [1739–40]).

8. The point is not, of course, that astrophysicists should be systematically uninterested in model systems as such; it is that the known properties of the particular system in question have (on the face of it) no relevant bearing on the questions and problems that interest astrophysicists.

9. David Kettler, "Historicism," in *The Blackwell Encyclopaedia of Political Thought*, ed. David Miller (Oxford: Blackwell, 1987), 208–11, offers a judicious discussion of the sometimes polemical use of the term by political theorists, noting, however, that in all versions, historicism rejects "an approach to the presuppositions and structure of knowledge grounded upon a universal theory of human nature, in favour of an approach grounded upon historically-localized knowledge of a historically changing world."

10. There are, of course, notable examples of intellectual historians (e.g., Quentin Skinner) taken seriously by political theorists. And some moral philosophers certainly do take history seriously; see, for example, Amartya Sen's well-known demonstration that famine does not occur in societies featuring constitutional democracies and freedom of information. And one could cite examples of the resort to history by moral philosophers with very different agendas, for example, "natural lawyers" such as John Finnis, who attempt to ground arguments against homosexuality on assertions that these constitute a consistent historical pattern of legal response to homosexual behavior by "decent societies."

11. Of course there are many important variants of each approach, and some very influential approaches that borrow from two or more of them; for example, John Rawls's influential work on justice seeks to undergird Kantian intuitions about rights with a social-contract thought experiment.

12. Hobbes translated Thucydides and regarded him as a staunch opponent of democracy. J. S. Mill lauded the massive Greek history written by George Grote, the first modern historian to find Athenian democracy frankly admirable. See Jennifer Tolbert Roberts, *Athens on Trial: The Antidemocratic Tradition in Western Thought* (Princeton: Princeton University Press, 1994), 246–47; Nadia Urbinati, *Mill on Democracy: From the Athenian Polis to Representative Government* (Chicago: University of Chicago Press, 2002).

13. Hobbesian social contract theory is an interesting case here in that Hobbes (like Jean-Jacques Rousseau and other social-contract philosophers) founds his conception of justice on the mythic pseudo-historical moment at which the voluntary

contract was struck between the sovereign and the subjects. The thought experiment that grounds the moral project is projected into a mythic past. Plato (*Republic*) and Aristotle (*Politics*, book 1) rely on similarly pseudo-historical myths of origin.

14. I recognize that this may be something of an overstatement, and I will note various examples of historical sensitivity on the part of political theorists in the notes below.

15. For example, Judith Jarvis Thomson's 1999 Tanner Lectures at Princeton, titled "Goodness and Moral Requirement," featured an extended hypothetical centered on the act of ringing a doorbell. I do not doubt that it was sound moral philosophy, but it is hard to find much meaningful connection between an argument about doorbells and the moral dilemmas confronted by most people.

16. This may be part of the reason that some of the most interesting contemporary North American political theory is concerned with matters of constitutional law. The legal issue, emerging from an actual problem that exists in complex historically situated human relations, here takes the place of the deracinated hypothetical. Other theorists have employed different means to try to get more real life into political theorizing, for example, by experimenting with dialogue format. See, for example, Daniel Bell, *Communitarianism and Its Critics* (Oxford: Oxford University Press, 1993).

17. I am assuming that "we" (the normative thinkers willing to poke about a bit in history imagined as the audience of this essay) share a particular history that inclines us to embrace certain normative values, that is, that we are heirs to a long and entrenched (although diversely interpreted) liberal tradition. Compare the historically situated arguments made by John Rawls, *Political Liberalism* (New York: Columbia University Press, 1996) to the earlier universalist stance of John Rawls, *A Theory of Justice* (Cambridge: Harvard University Press, 1971). Those who, for example, embrace a set of values centered on firm social hierarchy, divinely ordained law, and unquestioning obedience to superiors, while rejecting the primacy of individual liberty, political equality, and the dignity of the person, will, I suppose, prefer to look for a very different sort of exemplary historical narrative.

18. Of course, we may care about many values other than these "core liberal" values.

19. I am assuming that it is the relative lack of empathy of the experimental biologist for individual mice and fruit flies that enables him or her to undertake various sorts of experimental work on them. Whether that lack of empathy (and the behavior it facilitates) is ethically justifiable has been the focus of recent philosophical studies of animal rights. The ancient historian escapes the experimental biologist's ethical dilemma in that, by definition, his or her research can have no affect on the well-being of the actual, long-departed human objects of his or her study.

20. Ober, *Athenian Revolution*, chap 2, explains in more detail what I mean by the terms "useful" and "meaningful."

21. See Roberts, *Athens on Trial*, passim.

22. For the record, I have often been identified with the latter of these. For critical responses to my approach, see the introductory notes to the individual chapters of Ober, *Athenian Revolution*.

23. For an example of refining the case itself, see Ober, *Athenian Revolution*, chap. 5, an attempt to sort out the events that led to the establishment of democracy, with special reference to demotic agency. For an example of political theoretical experimenting with the case, see Josiah Ober, "Quasi-rights: Participatory Citizenship and Negative Liberties in Democratic Athens," *Social Philosophy and Policy* 17 (2000): 27–61, an attempt to clarify the nature of rights by contrasting contemporary conceptions of inherent rights with the Athenian conception of pragmatically established and maintained immunities.

24. Of course, these are not the only ways that Athenian democracy differs from a modern liberal regime. Athens was also a slave-holding society that limited political rights to men. See further below.

25. For example, see Derek L. Phillips, *Looking Backward: A Critical Appraisal of Communitarian Thought* (Princeton: Princeton University Press, 1993), 129–48.

26. These methodological assumptions inform the projects that led to two collections juxtaposing essays by theorists and ancient historians: J. Peter Euben, John Wallach, and Josiah Ober, eds., *Athenian Political Thought and the Reconstruction of American Democracy* (Ithaca: Cornell University Press, 1994); and Josiah Ober and Charles W. Hedrick, eds., *Demokratia: A Conversation on Democracies, Ancient and Modern* (Princeton: Princeton University Press, 1996).

27. Barry S. Strauss, *Athens after the Peloponnesian War: Class, Faction and Policy, 403–386 B.C.* (London: Croom Helm, 1986), with literature cited.

28. Michael Walzer, *The Company of Critics: Social Criticism and Political Commitment in the Twentieth Century* (New York: Basic Books, 1988). Note that Walzer's approach is to treat actual historical cases of political critics.

29. Josiah Ober, *Political Dissent in Democratic Athens: Intellectual Critics of Popular Rule* (Princeton: Princeton University Press, 1998).

30. On the ancient concept of rights, see Martin Ostwald, "Shares and Rights: 'Citizenship' Greek Style and American Style," in Ober and Hedrick, *Demokratia*, 49–61. In contrast, see F. D. Miller, *Nature, Justice, and Rights in Aristotle's Politics* (Oxford: Oxford University Press, 1995).

31. Ober, "Quasi-rights."

32. Stephen Macedo, *Diversity and Distrust: Civic Education in a Multicultural Democracy* (Cambridge: Harvard University Press, 2000).

33. Josiah Ober, "The Debate over Civic Education in Classical Athens: Education in Greek and Roman Antiquity," in *Education in Greek and Roman Antiquity*, ed. Y. L. Too (Leiden, the Netherlands: Brill, 2001), 273–305.

34. For the idea of Athens as built on oppression, see J. Roberts, "Athenian Equality: A Constant Surrounded by Flux," in Ober and Hedrick, *Demokratia*, 187–202.

Contrast: Edward E. Cohen, *The Athenian Nation* (Princeton: Princeton University Press, 2000).

35. See Ober, *Athenian Revolution*, chap. 2.

36. And hence Isaiah Berlin's notion of the only genuine good as a commitment to a diversity of goods. Perhaps it is not surprising that his work in ethical theory is based on intellectual history: embracing history forms a central part of Berlin's project, even while he remains horrified at the excesses that emerge from history. See further John Gray, *Isaiah Berlin* (Princeton: Princeton University Press, 1996).

Latitude, Slaves, and the Bible:
An Experiment in Microhistory

CARLO GINZBURG

My piece could take as its motto Mies van der Rohe's famous words "less is more." By knowing less, by narrowing the scope of our inquiry, we hope to understand more. This cognitive shift has been compared to the dilation and constriction of a camera lens.[1] One might call this approach microhistory, but labels are ultimately irrelevant.

1. My approach to microhistory has been largely inspired by the work of Erich Auerbach, the great Jewish scholar who spent his most creative years in Istanbul in exile from Nazi Germany. At the end of his masterpiece, *Mimesis*, written in Istanbul during the Second World War, Auerbach wrote: "Beneath the conflicts, and also through them, an economic and cultural leveling process is taking place. It is still a long way to a common life of mankind on earth, but the goal begins to be visible."[2]

Half a century later one hesitates to describe the so-called globalization that is taking place under our eyes as an "economic leveling process." On the other hand, the "cultural leveling," the erasure of cultural specificities that Auerbach looked at with growing worry, has become an unquestionable reality, although one difficult to grasp. In an essay published in 1952, Auerbach remarked that Goethe's concept of *Weltliteratur* had become increasingly inadequate to our endlessly expanding gaze. How can a philologist from a single cultural tradition approach a world in which so many languages and so many cultural traditions interact? Auerbach believed that one has to look for *Ansatzpunkte*, for starting points, for concrete details from which the global process can be inductively reconstructed.[3] The ongoing unification of the world, Auerbach wrote in the conclusion of *Mimesis*, "is most concretely visible now in the unprejudiced, precise, interior and exterior representation of the random moment in the lives of different people."[4]

Auerbach's strategy, the collecting and elaborating of *Ansatzpunkte*, was based on the cognitive model he had previously detected in the work of Marcel Proust and Virginia Woolf.[5]

2. I will return to this symmetry later. Some time ago, while I was working on a separate project, I came across a tract bearing the following title: *Mémoire sur le Pais des Cafres, et la Terre de Nuyts, par raport à l'utilité que la Compagnie des Indes Orientales en pourroit retirer pour son commerce* (Remarks on Kafirland and the Land of Nuyts, considered from the point of view of their usefulness to the trade of the East India Company). The copy I consulted at the University of California, Los Angeles Research Library—a photocopy of the original edition—is bound with a *Second Mémoire sur le Pais des Cafres, et la Terre de Nuyts*, also issued in Amsterdam in 1718. At the end of the two tracts, the identity of the author is revealed: Jean-Pierre Purry, a name I had never heard before. After a glance at the two texts I was immediately intrigued, for reasons that I shall discuss later. Then began a research project that is still far from its conclusion. This essay is a preliminary report on my work in progress.

3. Jean-Pierre Purry was born into a Calvinist family in Neuchâtel in 1675.[6] His father, Henry, a tinsmith (like his father and grandfather), died when Jean-Pierre was one year old. The following year, Henry's widow, Marie Hersler, bettered her lot by marrying the well-to-do Louis Quinche. While in his late teens, Jean-Pierre was appointed tax collector of Boudry, a little town near Neuchâtel; after one year, for unknown reasons, he gave up his post. On September 26, 1695, Jean-Pierre married Lucrèce Chaillet, the daughter of Charles Chaillet, the pastor of Serrières. Between 1696 and 1710 eight children were born to the couple; four of them died at an early age.[7] In 1709, Jean-Pierre was appointed mayor of Lignières.[8] Two years later, his precocious political career ended abruptly when he was compelled to resign the mayorship. Personal misfortunes were mentioned: a fire had damaged his house; a two-year-old venture selling wine to England had ended in financial disaster.

Given that for two thousand years the slopes surrounding Lake Neuchâtel had been covered with vineyards, Jean-Pierre's involvement in the wine trade is not surprising. Neither is the support he received during that crisis from his family and his wife's; after all, three marriages connected the Purrys to the Chaillets.[9] Yet in hindsight these events take on a remarkable singularity, the lineaments of a destiny. Jean-Pierre Purry's life would unfold itself under a constellation whose defining stars were wine, England, and a propensity to take great risks followed by great failures.

4. By the time the people of Lignières learned that their mayor had stepped down, Purry had left his birthplace behind and set out into a far larger world.[10] On May 26, 1713, he embarked as a corporal on a ship owned by the Dutch East

India Company, the instrument of Dutch economic and political expansion in Southeast Asia. In his position as the leader of seventy men, Purry may have had some knowledge of Dutch. The ship made a halt in Capetown and reached Batavia on February 2, 1714. Purry was to spend four years there, working as an employee of the Dutch East India Company. On December 11, 1717, he left Batavia, embarked as an accountant. After the usual halt in Capetown, his ship reached the Netherlands on July 17.[11]

These factual data provide the context for the writings from which I started, Purry's two *Mémoires sur le Pais de Cafres et la Terre de Nuyts*. Let us now take a closer look at them.

5. In the first tract, addressed to the Assembly of the Seventeen, the board that led the Dutch East India Company, Purry tried to convince the governor of the company either to colonize Kafirland (today's South Africa) or, alternatively, the Land of Nuyts (today's western coast of Australia).[12] In his second *Mémoire*, dated September 1, 1718, well after his return to Europe, Purry replied to the objections raised by his opponents and made a strong case for the colonization of the Land of Nuyts.

Purry's projects were rooted in a theory about climate, which he explained at length in his first *Mémoire*. He rejected labels like "temperate" or "cold" as exceedingly vague, and as absurd the standard praise showered on France's geographical location in the middle of the temperate zone, between forty-two and fifty-one degrees of latitude. The grapes that grew at fifty-one degrees of latitude, he objected, produced undrinkable wine, after all. The best climate in the world was found at thirty-three degrees of latitude.

Purry's theory was that of a former wine merchant, born in a region noted for its wines. But his seemingly superficial remarks had more complex implications. He provided a list of countries located between thirty and thirty-six degrees of latitude: Barbary, Syria, Chaldea, Candia, Cyprus, Persia, Mongolia, "the middle part of China," Japan. But, he explained, those that are closer to thirty-three degrees of latitude "far surpass the others in fertility, as one can see even in the land of Canaan, of whose provinces Galilee is one of the finest."[13]

This passing and underplayed allusion ("even [même]") was a crucial reference to the biblical book of Numbers, chapter 13, and it gave Purry's argument a sudden twist. Let us make explicit the biblical reference, which Purry quoted in full in his second *Mémoire*.

6. "And the Lord spake unto Moses saying, Send thou men, that they may search the land of Canaan, which I give unto the children of Israel." Obeying the com-

mand, Moses sent men from each tribe of Israel "to spy out the land of Canaan, and said unto them, Get you up this *way* southward, and go up to the mountain: and see the land, what it is. . . . And bring the fruit of the land. Now the time *was* the time of the firstripe grapes." The spies come to Hebron, and then "unto the brook of Eshcol, and cut down from thence a branch with one cluster of grapes, and they bare it between two upon a staff" (Num. 13:1–2, 17–18, 23).

Grapes and wine, once again. The enormous bunch of grapes brought by two men on a staff symbolized the extraordinary richness of the Promised Land. Thanks to the reference to Canaan, the hidden core of Purry's project emerges.[14] There are two basic types of quotations in his two *Mémoires*. On the one hand, seventeen references to the Old Testament (plus two implicit allusions to it), as well as a single quotation from Paul's first letter to the Corinthians; on the other, fifteen allusions to contemporary geographical and historical accounts. But the biblical references provide a key to the secular passages. The perfect latitude was, first of all, the latitude of the Promised Land. Purry's plans for colonial settlement were based on the biblical Exodus—although his reading of the Bible was, as we will see, flexible enough to allow him, for instance, to look for the perfect latitude of thirty-three degrees in both the boreal and the austral hemispheres.

7. The long-term impact of the Exodus narrative is well known. Many years ago Michael Walzer argued that the journey of the children of Israel from slavery to freedom, from Egypt to the Promised Land, had provided, down through the centuries, a revolutionary model devoid of messianic connotations, which inspired—as Walzer remarked echoing Gershom Scholem—the modern Zionist movement.[15] But those revolutionary interpretations, Walzer admitted, ignore a section of the Exodus narrative: the conquest, the war against the Canaanites who inhabited the land. As he rejected the reading of the Exodus offered by right-wing Zionists, Walzer implicitly aligned himself with the motto of liberal Zionism: "A people without land [the Jews] found a land without people [Palestine]." In this reading, Canaanites are silently deleted from the biblical narrative; likewise, Palestinians have been bracketed from the official version of Israeli history, which over the past decades has been the target of a new generation of Israeli historians.[16] On a more general hermeneutic level, two questions come to mind. First, is it allowed to bracket the conquest of Canaan from the biblical narrative simply because one dislikes the way in which that conquest has been symbolically used in contemporary political debates? Second, is such bracketing compatible with Walzer's principle (also derived from Scholem and

certainly open to discussion) that the meaning of the biblical narrative ultimately coincides with the full range of its interpretations?[17]

Purry implicitly regarded both the Canaanites and the war waged against
them by the children of Israel as a crucial feature of the biblical narrative. In his
reading, the journey toward the Promised Land became a model and a justification for the European conquest of the world.[18]

8. Purry tried to convince the Dutch East India Company to send immigrants
either to South Africa or to Australia. But the relatively small number of Europeans likely to immigrate to that area drove him to consider a different alternative: "When one is unable even to find laborers, one can have slaves work the
soil. The Romans did not work their own soils otherwise."[19]

Why did Purry justify slavery with a secular precedent instead of quoting,
as he usually did, a passage from the Old Testament? Possibly because the curse
Noah set on the children of Ham, who had seen his nakedness, seemed to connect slavery to an inborn stigma.[20] Purry's attitude was different. He dismissed
the idea that slaves had limited learning capacities. In Java he had seen slaves of
both sexes working as tailors, carpenters, and shoemakers. They played musi-
cal instruments at weddings; they danced. Those things "are nothing but the
effects of habit and continual practice. I can see, as a result, no reason why
slaves should be incapable of learning the science of agriculture." At this point
an imaginary opponent suggested a graver impediment: "It will be objected that
in this case justice and equity will bar us from setting ourselves up in the Land
of Nuyts and lording it over those who have been there, father and son, for as
long as several thousand years, and will also bar us from evicting from their
land people who have never done us any harm."[21]

9. Here was a striking and quite straightforward objection to European colonization as such. An even more striking rebuttal followed. There was no injustice
in this, Purry replied, for two reasons. First of all, "the Earth belongs to God in
perpetuity, and we have but the use of it, something like the father who has a
dish set before his children or his servants: he does not assign a portion to each,
but rather that which each fairly seizes for himself belongs to him, though before that he had no greater right to it than the others, and though they did not
grant him permission to take this or that piece."[22]

A large family meeting around the table, children or servants cheerfully
trying to grasp their meal. This patriarchal scene was an implicit comment on
Leviticus, a biblical passage quoted by Purry: "The land shall be not sold for

ever: for the land *is* mine; for ye *are* strangers and sojourners with me" (Lev. 25:23).

Only recently I realized that Purry's words had been also silently inspired by a different text: John Locke's *Second Treatise of Government*. "In the beginning all the World was *America*," Locke wrote; "God gave the World to Men in Common." But property, being based on industry, was legitimate; otherwise, Locke argued, if "an explicit consent of every Commoner" would be "necessary to any ones appropriating to himself any part of what is given in common, Children or Servants could not cut the Meat which their Father or Master had provided for them in common, without assigning to every one his peculiar part." Purry must have read Locke's *Second Treatise* in David Mazel's French translation, published in Amsterdam in 1691, the year after the first English edition, and then often reprinted.[23]

Purry gave an original twist to Locke's reflections. "Given that all men," Purry went on,

> naturally possess the same rights over the goods of the World, thanks to the intention of the creator, who bestowed on them this commonly held right in order that they might exercise it, it does not seem reasonable that the simple state of possession, albeit thousands of years old, should privilege the claims of any individual over the others, without their consent, which is to say, without some agreement they have made between them on the subject. And as long as each person takes only what he needs, he is not infringing on the rights of others, who could, in turn, claim the privilege of the first occupant after a fashion.[24]

Purry here answered an implicit question: was the European conquest of the world legally justified? To raise such a question already implied a distancing, if not perhaps a doubt. Purry articulated his answer in terms of a natural law, which he derived from a biblical passage, although one could also argue the opposite, that a notion of natural law inspired by Locke's *Second Treatise* inspired Purry's reading of the Bible.[25] Locke's passage on the human bond connecting "a *Swiss* and an *Indian*, in the Woods of *America*" who are "perfectly in a State of Nature, in reference to one another" must have had a special resonance for Purry, a Swiss.[26] Before God there were no hierarchies; every human being had the same right to use the earth. Local bonds were nullified by the invocation of God, a God distant and lonely in his uniqueness. Claims rooted in antiquity, in traditions thousands of years old, had no validity whatsoever. No property could be held in perpetuity; only the present counted. The earth was like a meal, and, in principle, everyone was entitled to get a share of it, but there would be no orderly distribution; in fact, there would be no distribution at all. In claiming a share, the children of God had to behave "fairly," of course, but the reference to the "rights of others" does not suggest a brotherly relationship. The "rights of others" refers to a law governing all; the biblical word *stranger* defined not only relationships between human beings and God but strictly human relationships as well. Everybody was a stranger to everybody else. This common shared condition did not, in Purry's global perspective, elicit the compassion that inspired Exodus 23:9: "Thou shalt not oppress a stranger: for ye know the heart of a stranger, seeing ye were strangers in the land of Egypt." When everybody can "in turn, claim the privilege of the first occupant *after a fashion*," when each individual is tacitly entitled "to take this or that piece," natural law turns (we might conclude) into a law of mutual pillage. Might becomes right. At this point, Purry's second axiom, and morality, are introduced:

> Savage and rustic people love above all things a lazy existence and . . . the more a people is simple and vulgar the less it is given to work, while a life of abundance and pleasure requires a great deal of care and trouble. In addition, the countries inhabited by these sorts of savage and lazy people are never very populous. Thus one has every reason to believe that far from harming the inhabitants of the Land of Nuyts—and one is not obliged to displace them—the establishment of a good European Colony would provide for them all sorts of benefits and advantages, as much because theirs would be a civilized life as because of the arts and sciences they would be taught.[27]

We are confronted with a series of overlapping, allegedly self-evident oppositions: civilized and savage life; industry and laziness; abundance and scarcity. The establishment of "a good European Colony" will rescue the savages from their sinful laziness and will provide for them "a civilized life."[28] The change brought by the Europeans will have been moral and profitable for everybody, "as long," Purry writes, "as one acts gently and regards them as poor creatures who, though vulgar and quite ignorant, are nonetheless members of human Society, as much as we are."[29]

Purry remarked that the Spaniards and the Portuguese, who treated the American Indians as if they were animals, had been despised for their cruelty and barbarity. His colonization projects, on the contrary, could be carried on "without causing the [local] inhabitants any suffering or in any way wronging them. These sorts of benefits, which never give rise to any regrets, and which may be conferred without in any way compromising one's decency and Christian spirit, are truly worthy of our Illustrious Company."[30]

To dismiss this kind of moral reasoning as either a mask concealing the features of greed, or as an out-and-out lie, would be simplistic. Purry's effort to eliminate regret was in itself significant. European colonization, at this stage and in certain environments, could generate bad conscience—a feeling to be silenced in the name of morality, civilization, and profit. The argument based on natural law that every human being stood equal before God and was equally amenable to civilization would contribute, in the long run, to antislavery and anticolonial movements of various kinds. But before that could happen, it would serve as an elaborate justification for European colonization.

10. Jean-Pierre Purry was accustomed to ocean crossings. He was born in Europe, spent some years in Asia, visited Africa, and ended his life in North America after having vainly championed the colonization of New Holland—today's Australia. Purry was able to view the earth as a whole. Not many individuals before him possessed so global and comprehensive a view; even fewer had the opportunity or the capacity to give written expression to what they saw and what they thought of it. How did Purry achieve this?

While it is clear that he was a fairly cultivated man, his educational background is unknown.[31] Above all, Purry thought with the Bible, an experience he shared with innumerable individuals before and after him.[32] The Bible gave him words, arguments, and stories; he projected words, experiences, and events into the Bible. Other books provided him with a lens through which to read the Bible and vice versa.

Let us consider a few examples. When objections were raised to Purry's plan to set up a large colony in South Africa, he scornfully rejected them: "Because to state that men are incapable of resolving to give up their connections, their friends, their relatives, these statements are foolishness and chimeras that one gets into one's head."[33]

To prove his point, Purry recalled in a single breath two quite different groups: the French immigrants to Canada, who spoke with regret "of the fine flavor of their melons, their partridges, and so many other things that make life delightful,"[34] and the children of Israel, who murmured against Moses and Aaron: "Would to God we had died by the hand of the Lord in the land of Egypt, when we sat by the flesh pots, *and* when we did eat bread to the full; for ye have brought us forth into this wilderness, to kill this whole assembly with hunger" (Exod. 16:3).

Purry explicitly sympathized with such a practical attitude. His passing allusion to "so many other things that make life delightful" sprang from a deep hostility toward all sorts of asceticism. To him, civilization meant abundance.

But here a contradiction in his mind emerges. On the one hand, he insisted that abundance could be had only through industry and hard work. On the other, he subscribed to the old myth of a land of easy, flowing abundance. What did men mean when they spoke of "a good country"? Purry asked during his discussion of the ideal latitude. Purry offered his own answer: "As for me, I feel that a good country is one that abounds not only in milk and honey, but generally in all the things that appeal to our sensuality and fill our lives with delights; a land of Cockaigne and gorgeous meals, fertile, producing easily, with little work and cheaply, all of life's necessities. This, briefly and according to my humble notions, is a good country."[35]

But Purry's antiasceticism and his praise of material goods were not relics from peasant utopias, as the allusion to the land of Cockaigne might suggest. Among the authors quoted in Purry's *Mémoires* one finds François Bernier—a professor of medicine at the University of Montpellier, a philosopher, and a traveler—and Sir William Temple—a politician, an essayist, and the patron of Jonathan Swift. Both Temple and Bernier (who knew each other) contributed to the reappraisal of Epicurus, the pagan philosopher—a major event in European intellectual history initiated by Pierre Gassendi in the mid-seventeenth century.[36] Following Epicurus's praise of pleasure, Temple depicted civilization, in his essay "Upon the Gardens of Epicurus" (1685), as the form of society beneficially ruled by ambition and avarice: a detached, ironical description famously developed in Bernard Mandeville's *Fable of Bees*. Temple's essays had a deep impact on Purry. One can see him pondering Temple's remark that "the best climate for the production of all sorts of the best fruits . . . seems to be from about twenty-five, to about thirty-five degrees of latitude."[37] Purry's reading of the Bible, filtered by Temple's essay and by geographical writings, led him to formulate his theory of the perfect latitude, located at thirty-three degrees.

11. Purry's projects were examined by the managers of the Dutch East India Company and ultimately rejected on April 17, 1719.[38] This is not surprising; the company preferred trade to colonization. More surprising is the fact that immediately after, in unknown circumstances, Purry became the director general of the French India Company.[39] By 1720 he was in Paris, fully immersed in the financial turmoil generated by John Law, the Scottish financier, and his "system." Purry invested the money he had earned in Batavia with some initial success.[40] According to a friend, Purry pursued a speculator's jackpot with utter determination, saying: "Here everybody speaks of millions. Once I'll have a few millions, I'll cash out."[41] The Mississippi Bubble popped, and Purry lost everything.

He gave up neither his theories nor his projects. On June 6, 1724, he wrote to Horatio Walpole asking to be introduced to the duke of Newcastle; Walpole promptly complied, the following day.[42] In a memorial addressed to the duke, published in London that same year, Purry proposed the colonization of South Carolina by several hundred Swiss Protestants. Frustrated in his designs on the austral hemisphere, Purry had shifted his focus to thirty-three degrees north latitude.

His first expedition to America ended in failure, and Purry returned destitute to his hometown.[43] He was confined by his family to a mountain farm not far from Neuchâtel. From there, Purry sent deferential letters to his stepbrothers requesting money for his little expenses: letters, tobacco. But, even here, he could not refrain from referring to his American projects.[44] Purry must have spent a number of years suspended between a miserable present and the expectation of a grandiose future. Then something happened. At last, official patronage came. On March 10, 1731, George II signed a royal patent authorizing Jean-Pierre Purry, a colonel in the British army, to found a city in South Carolina, to be named Purrysburg. As Purry had proposed, it was to be inhabited by a settlement of Swiss Protestants.[45]

An advertising campaign mounted by Purry must have contributed to the flow of immigrants to his colony. Purry's detailed descriptions of South Carolina were published in Switzerland and translated into German and English.[46] In the *Eclaircissemens* appended to a second edition of his pamphlet, published in 1732, Purry responded to the murmurs of newly arrived colonists. Somebody, for instance, had complained about the region's climate. Purry, who was always willing to air his theories about latitude, adopted a startled tone: "To say that the region of Carolina is too hot for the Europeans, and especially for the Swiss, is as absurd as it would be to complain about Syria, or, as it was formerly known, the Land of Canaan."[47]

Like Moses (a metaphor he would have liked), Jean-Pierre Purry was not allowed to see the promised land of industrial revolution. He died on August 18, 1736, in the city bearing his name.[48] The city itself decayed and ultimately disappeared. Jean-Pierre's eldest son, Charles, was murdered in a slave revolt in 1754. Another son, David, who had stayed in Europe, became enormously rich. At his death, in 1786, he left his money, part of which had been earned through the slave trade with Brazil, to the poor people of Neuchâtel. His statue is placed in the middle of the city main square, which bears his name.

12. Jean-Pierre Purry's colorful life certainly deserves a detailed reconstruction. One could tell a story, even a good story about him. But the aim of my project

is different. Since the very beginning of my research I have tried to answer the following question: can an individual case, if explored in depth, be theoretically relevant?

When I first looked at Purry's two *Mémoires*, I immediately thought of Max Weber's *The Protestant Ethic and the Spirit of Capitalism*. In that famous essay, first published in 1904–5, Weber argued that the emergence of an attitude he called "inner-worldly asceticism" (*innerweltliche Askese*), inspired by Calvinism and its Puritan developments, played a crucial role in the emergence of capitalism by submitting economic activity to rational control.[49] Weber's controversial thesis, which has been debated ever since, focused on entrepreneurs as agents of change, stressing the psychological impact of religious concepts like calling (*Beruf*). But, as it has been noticed, individual entrepreneurs affected by Protestant ideas are, surprisingly enough, absent from Weber's essay. Benjamin Franklin, whose reflections Weber repeatedly quoted, is a late and rather secularized case.[50] Jean-Pierre Purry seems on the contrary a perfect illustration of Weber's thesis: a Calvinist entrepreneur, fully committed to the Protestant cause, extensively quoting the Bible to argue his colonization plans, and shaping his own life according to a geographic theory centered on the Land of Canaan. But, as soon as my research really began, its aim became less obvious.

As I immediately realized, to prove or to disprove Weber's argument was beside the point. On the one hand, Weber never argued his case as a clear-cut, clearly disprovable statement like "all swans are black." A white swan, a non-Calvinist entrepreneur, obviously did not affect Weber's argument at all. On the other, a Calvinist entrepreneur like Purry could never prove an argument like Weber's, which had been formulated in an abstract, ideal-typical form. As Weber repeatedly stressed, "to speak in terms of ideal-types [*Ideal-typen*]" means, "in a certain sense," to do "violence to historical reality."[51] Like Plato's ideas, ideal types are immune to contradictions.[52] According to Weber's definition, "an historical individual [is] a complex of elements associated in historical reality which we unite into a conceptual whole from the standpoint of their cultural significance."[53] A human being is of course a more unpredictable, not to say contradictory, reality. The gap between Jean-Pierre Purry and the ideal-typical Calvinist entrepreneur is part of Weber's postulates. But Weber himself repeatedly stressed that ideal-typical constructions must be continuously submitted to the test of empirical research. What can be the result of a test based on Purry's case?

Besides the convergences I already mentioned, some areas of equally obvious divergence come out: Purry's antiasceticism and his justification, based on his own reading of the Bible, and especially of the Exodus narrative, of Europe's conquest (including slavery and the use of force) of the world. The second point throws some interesting light over the genesis and meaning of Weber's *The Protestant Ethic and the Spirit of Capitalism*. Many readers regarded it as an argument against Marxism, positing a religious cause to capitalism rather than an economic one. Weber strongly objected that his aim had not been "to substitute for one-sided materialistic an equally one-sided spiritualistic causal interpretation of culture and of history."[54] Weber's polemical relationship with Marx was indeed subtler and closer. I would argue that Weber's *The Protestant Ethic and the Spirit of Capitalism* was written not only *against* the section of Marx's *Capital* that starts with chapter 26, "The Secret of Primitive Accumulation," but also *with* it, reassembling and turning upside down some of its passages.[55]

Marx's discussion opens with the following sentence: "This primitive accumulation plays approximately the same role in political economy as original sin does in theology." According to this "theological" version, "long ago there were two sorts of people; one, the diligent, intelligent, and above all frugal élite; the other, lazy rascals, spending their substance, and more, in riotous living. . . . In actual history," Marx goes on, "it is a notorious fact that conquest, enslavement, robbery, murder, in short, force, play the greatest part."[56]

In a sense, Weber consciously elaborated a range of subtle arguments to support the "theological" interpretation of primitive accumulation. On the one hand, he stressed the role of ascetic frugality in capitalist ethic; on the other, he traced a firm boundary between capitalistic adventurers and genuine capitalist entrepreneurs. Capitalistic adventurers "existed everywhere," in all sort of societies: a curious remark, hardly compatible with the gloss that "in overseas policy they have functioned as colonial entrepreneurs, as planters with slaves, or directly or indirectly forced labour."[57] The last issue was crucial. In Weber's view, genuine capitalist entrepreneurs had nothing to do with force.[58]

Marx, on the contrary, pointed at the role played by the colonies in the process of primitive accumulation: "The veiled slavery of the wage-labourers in Europe needed the unqualified slavery of the New World as its pedestal."[59] After having recalled the frightening treatment of indigenous populations in plantation colonies, Marx noticed that "even in the colonies properly so called, the Christian character of primitive accumulation was not belied." This claim was illustrated as follows: "In 1703 those sober exponents of Protestantism, the Puritans of New England, by decrees of their assembly set a premium of £40 on every Indian scalp and every captured redskin"; in 1744, "for a male scalp of 12 years and upwards, £100 in new currency; for a male prisoner £105, for women and children prisoners £50, for the scalps of women and children £50."[60]

To put such chilling punctiliousness under the rubric "Christian character of primitive accumulation" is typical of Marx's sarcasm. In the same mode, he evoked "the 'spirit' of Protestantism" to describe the introduction of merciless, punctilious poor laws into Elizabethan England.[61] But in Weber's use of the "spirit of capitalism" (a "somewhat pretentious phrase," he admitted), there is not the faintest trace of irony. His attempt to demonstrate the Christian (more specifically, Calvinist) character of primitive accumulation was equally serious. Marx's ferocious remarks were turned upside down and became the starting point of Weber's essay. But, when he praised "exact calculation" as a feature of rational capitalistic organization,[62] Weber probably did not recall the Puritan calculation of redskin scalps.

The model Weber advanced in *The Protestant Ethic and the Spirit of Capitalism* by systematically erasing violence from the early history of capitalism is greatly inferior to Marx's. On the other hand, Weber was certainly right in focusing on the role played by agents influenced by religion—a crucial issue that Marx ignored. But which agents? Purry, the Protestant entrepreneur who stressed the necessity of force to bring lazy, uncivilized natives to the realm of abundance, is incompatible with Weber's ideal type. If I am not mistaken, Purry's case compels us to reconsider from a sharply focused, unexpected angle the comparative strengths and weaknesses of the two most influential social thinkers of our time.

13. My approach to microhistory is strongly indebted to the work of scholars like Erich Auerbach (whom I mentioned earlier) who developed interpretations of literary artifacts based on clues others had considered insignificant. This version of microhistory has been contrasted with another version, one more oriented toward the social sciences and the critique of their methods.[63] In my view, the opposition is groundless because both versions of microhistory aim at the same theoretical target, albeit from opposite directions. I know that the word cannot be taken for granted in this context. In the social sciences, theory is often tacitly identified with a broad approach à la Max Weber, and microhistory with a narrowly focused attempt to rescue from oblivion the lives of marginal, defeated people. If one accepts these definitions, microhistory would be confined to a peripheral and basically atheoretical role that leaves the dominant theories unchallenged. The case of Jean-Pierre Purry, that early prophet of the capitalist conquest of the world, stands a chance of knocking down some of the barriers thought to divide microhistory and theory.[64] A life chosen at random can make concretely visible the attempt to unify the world, as well as some of its implications.

In saying this I am echoing Auerbach. But Auerbach was implicitly referring to Proust. Let us allow Proust to have the final word: "People foolishly imagine that the broad generalities of social phenomena afford an excellent opportunity to penetrate further into the human soul; they ought, on the contrary, to realise that it is by plumbing the depths of a single personality that they might have a chance of understanding those phenomena."[65]

NOTES

This article was first published in English in *Critical Inquiry* 31 (2005): 665–84. Reprinted by permission of The University of Chicago Press. Different versions of the essay were delivered in Istanbul (see Carlo Ginzburg, "Küreselleşmeye Yerel Bir Yaklasim: Coğrafya, Köleler ve İncil," in *Tarih Yaziminda Yeni Yaklaşimlar: Küreselleşme ve Yerelleşme*, 2 vols. [Istanbul: Tarih Vakfı, 2000], 1:17–39); and at the University of California, Los Angeles; Central European University, Budapest; the University of Pennsylvania; Boston University; the University of Oslo; the University of São Paulo; Columbia University; the Facoltà di Letttere e Filosofia, Siena; the Université Libre, Brussels; Rossiskii Gosudarstvennyi Gumanitarnyi Universitet, Moscow (see *Sciroty, raby, i Biblia: Opit mikroistorii* [Moskow: Rossiskii Gosudarstvennyi Gumanitarnyi Universitet, 2003]); and at the University of Chicago. The essay was also given as the annual Nexus Lecture 2002, organized by the Nexus Institute in Tilburg, the Netherlands; it was originally published as Carlo Ginzburg, "Geografische breedte, slaven en de Bijbel: Een experiment in microgeschiedenis," *Nexus* 35 (2003): 167–84, and appears here with their permission. Many thanks to Carlo Aguirre Rojas, Perry Anderson, Pier Cesare Bori, Alberto Gajano, Stefano Levi Della Torre, and Marta Petrusewicz for having helped me, either directly or indirectly, with their comments and suggestions; to Albert de Pury for his generous help; and to Samuel Gilbert for his stylistic suggestions.

1. See Jacques Revel, introduction to *Le pouvoir au village: Histoire d'un exorciste dans le Piémont du XVIIe siècle*, by Giovanni Levi, trans. Monique Aymard (Paris: Gallimard, 1989), i–xxxiii, xv.

2. Erich Auerbach, *Mimesis: The Representation of Reality in Western Literature*, trans. Willard R. Trask (Princeton: Princeton University Press, 1953), 552.

3. See Erich Auerbach, "Philology and *Weltliteratur*," *Centennial Review* 13 (1969): 1–17.

4. Auerbach, *Mimesis*, 552.

5. The parallel is explicitly drawn a few pages before; see Auerbach, *Mimesis*, 548.

6. See L.-E. Roulet, "Jean-Pierre Purry et ses projets de colonies en Afrique du Sud et en Australie," *Musée Neuchâtelois* 31 (1994): 49–63; and L.-E. Roulet, "Jean-Pierre Purry explorateur (1675–1736)," in *De Saint Guillaume à la fin des Lumières*, vol. 1 of *Biographies Neuchâteleoises*, ed. Michel Schlup (Neuchâtel-Hauterive, Switzer-

land: G. Attinger, 1996), 237–42; Arlin C. Migliazzo, "A Tarnished Legacy Revisited: Jean Pierre Purry and the Settlement of a Southern Frontier, 1718–1736," *South Carolina Historical Magazine* 92 (1991): 232–52; and Arlin C. Migliazzo, ed., *Lands of True and Certain Bounty: The Geographical Theories and Colonization Strategies of Jean Pierre Purry*, trans. Pierette C. Christianne-Lovrien and 'BioDun J. Ogundayo (Selinsgrove, Pa.: Susquehanna University Press, 2002). See also H. Jéquier, Jacques Henriod, and Monique de Pury, *La famille Pury* (Neuchâtel, Switzerland: Caisse de Famille Pury, 1972). None of these studies analyzes the religious arguments for colonization put forward by Purry. The spelling of the family name varies (Purry, Pury, Puri, Purri); see *Recueil de quelques lettres et documents inédits concernant David de Pury et sa famille* (Neuchâtel, Switzerland: H. Wolfrath, 1893), 11n1. I chose Purry, the version Jean-Pierre consistently used.

7. See *Recueil de quelques lettres et documents inédits concernant David de Pury et sa famille*, 73–75.

8. See Archives de l'État, Neuchâtel, Switzerland, Archives de la famille de Pury, G. XII; see also Roulet, "Jean-Pierre Purry et ses projets de colonies en Afrique du Sud et en Australie," 51.

9. See *Recueil de quelques lettres et documents inédits concernant David de Pury et sa famille*, 8.

10. "Leur curiosité naturelle les porte [Neuchâtel's inhabitants] la plûpart à voïager dans les païs étrangers" (D. F. de Merveilleux, *La parfaite introduction à la géographie universelle*, 2 vols. [Neuchâtel, Switzerland: J. J. Schmid, 1690], 2:515).

11. On this, I follow Roulet, "Jean-Pierre Purry et ses projets de colonies en Afrique du Sud et en Australie," based on a lecture given by C. C. Macknight in 1993. I am very grateful to Albert de Pury, who sent me a typewritten version of Macknight's unpublished lecture.

12. The land was named for Pieter Nuyts, extraordinary councillor of India, who discovered it in 1627; see J. E. Heeres, *Het Aandeel der Nederlanders in de Ontdekking van Australie, 1606–1765* (Leiden, Netherlands: E. J. Brill, 1899), 51.

13. "Surpassent de beaucoup la fertilité des autres, ainsi qu'on peut remarquer même au païs de Canaan, dont la Galilée étoit l'une des meilleures provinces" (Jean-Pierre Purry, *Mémoire sur le Pais des Cafres, et la Terre des Nuyts, par raport à l'utilité que la Compagnie des Indes Orientales en pourroit retirer pour son commerce* [Amsterdam: Humbert, 1718], 17–18; hereafter abbreviated as *M*).

14. See Roulet, "Jean-Pierre Purry et ses projets de colonies en Afrique du Sud et en Australie," 55.

15. See Michael Walzer, *Exodus and Revolution* (New York: Basic Books, 1985), 123.

16. See Edward W. Said, "Michael Walzer's 'Exodus and Revolution': A Canaanite Reading," *Grand Street* 5 (1986): 86–106. A further exchange between Said and Walzer has been republished by William D. Hart, *Edward Said and the Religious Effects of Culture* (Cambridge: Cambridge University Press, 2000), 187–99 (kindly brought to my attention by David Landes).

17. See Walzer, *Exodus and Revolution*, 7–8. A comparison between this Jewish theme

and the Christian hermeneutic tradition, brilliantly analyzed by Pier Cesare Bori, *L'interpretazione infinita: l'ermeneutica cristiana antica e le sue transformazioni* (Bologna: Il Mulino, 1987), would be rewarding.

18. "Davanti al Muro capivo perché la leggenda americana, quella della frontiera e dei massacratori di indiani, si fosse nutrita del libro dell'Esodo" (Franco Fortini, *Extrema ratio: Note per un buon uso delle rovine* [Milan: Garzanti, 1990], 67).

19. "Quand même on ne trouveroit point de laboureurs, on pourroit en ce cas là faire cultiver la terre par des esclaves. Les Romains ne labouroient pas les leurs autrement" (*M*, 69).

20. See Robin Blackburn, *The Making of New World Slavery: From the Baroque to the Modern, 1492–1800* (New York: Verso, 1997), 64–76, on Noah's curse, with an extensive bibliography.

21. "Ne sont autre chose que des effets de l'habitude et d'une exercice continuel. Ainsi je ne voy pas pourquoi des esclaves ne pourroient pas apprendre la science de l'agriculture" (*M*, 69–70); "Mais, dira-t-on, quand cela seroit, la justice ni l'équité ne permettent pourtant pas qu'on pût s'aller établir dans la Terre de Nuyts au prejudice de ceux qui y sont déjà de pere en fils, depuis, peut-être, quelque milliers d'années, ni qu'on pût chasser de leur païs des gens qui ne nous ont jamais fait aucun mal" (*M*, 70–71).

22. "La terre apartient toûjours à Dieu en toute proprieté, et nous n'en avons que l'usufruit, à peu près de même qu'un pere de famille qui fait servir quelque plat à ses enfans ou à ses domestiques, il n'assigne pas à chacun sa portion, mais ce dont chacun se saisit honnêtement est à lui, quoi qu'auparavant il n'y eût pas plus de droit que les autres; et quoi que ceux ci ne lui aient pas donné la permission de prendre tel ou tel morceau" (*M*, 71).

23. John Locke, *Second Treatise of Government*, 1689, in *Two Treatises of Government*, ed. Peter Laslett (Cambridge: Cambridge University Press, 1963), 319, 309, 307. For the French translation, see John Locke, *Du gouvernement civil où l'on traitte de l'origine, des fondemens, de la nature, du pouvoir et des fins des sociétés politiques, traduit de l'anglois*, trans. David Mazel (Amsterdam: Chez Abraham Wolfgang, prés de la Bourse, 1691).

24. "Tous les hommes ayant donc naturellement le même droit sur les biens du Monde en vertu de l'intention du createur qui ne leur a donné ce droit commun qu'afin qu'ils en fissent usage, on ne conçoit pas qu'une simple possession, quoique de plusieurs milliers d'annéés, puisse être valable en faveur de quelqu'un à préjudice des autres, sans le consentement de ceux-ci, c'est-à-dire, sans quelque convention faite entr'eux à ce sujet: et tant que chacun ne prend que ce qu'il lui faut, il ne donne aucune atteinte au droit des autres, qui peuvent à leur tour, faire valoir d'une maniere ou d'autre, le privilège du premier occupant" (*M*, 70–71).

25. See ibid. The central role played by natural right in Purry's argument emerges again in *Second Mémoire sur le Pais des Cafres, et la Terre de Nuyts* (Amsterdam: P. Humbert, 1718), 52; hereafter abbreviated as *SM*: "Mais je suis très persuadé qu'on peut presque se promettre d'avance le succès d'une bonne entreprise,

lorsqu'elle n'a rien de contraire au droit naturel, et que le Ciel ne manque jamais d'accompagner de ses benédictions des desseins qui sont fondés sur la Charité envers le prochain, aussi bien que sur l'Amour de Dieu." On this issue, see Anthony Pagden, *The Fall of Natural Man: The American Indian and the Origins of Comparative Ethnology* (Cambridge: Cambridge University Press, 1982).

26. Locke, *Second Treatise of Government*, 295.

27. "Les gens sauvages et rustiques aiment la vie faineante par dessus toutes choses, et . . . plus un peuple est simple et grossier moins il est adonné au travail: au lieu qu'une vie d'abondance et de delices demande beaucoup de soins et de peine. Ajoûtons à cela, que les païs qui sont habités par ces sortes de gens sauvages et paresseux ne sont jamais fort peuplés. Ainsi on a tout lieu de croire, que bien loin de causer du dommage aux habitans de la Terre de Nuyts, ni qu'on fût obligé de les chasser chez eux, au contraire, l'établissement d'une bonne Colonie Européenne leur procureroit toutes sortes de biens et d'avantages, tant pour une vie civilisée que par les arts et les sçiences qu'on leur enseigneroit" (*M*, 72–73).

28. Lucien Febvre, "Civiltà: evoluzione di un termine e d'un gruppo di idee," *Studi su Riforma e Rinascimento: e altri scritti su problemi di metodo e di geografia storica* trans. Corrado Vivanti (Turin: G. Einaudi, 1966), 5–45, 15n1 quotes Antoine Furetière's *Dictionnaire universel, contenant généralement tous les mots françois tant vieux que modernes, et les termes de toutes les sciences et des arts . . .* (The Hague: Arnout et Renier Leers, 1690); "La prédication de l'Évangile a civilisé les peuples barbares les plus sauvages." See also Émile Benveniste, "Civilisation: Contribution à l'histoire du mot," in *Hommage à Lucien Febvre. Éventail de l'histoire vivante offert par l'amitié d'historiens, linguistes, géographes, économistes, sociologues, ethnologues*, 2 vols. (Paris: A. Colin, 1953), 1:47–54.

29. *M*, 72–73.

30. "Sans aprehender de faire souffrir ses habitans, ni de commettre aucune injustice à leur égard. De tels biens, qui ne donnent jamais aucun remord et qu'on peut acquerir sans donner la moindre atteinte à la qualité d'honnête homme et de Chrétien, sont véritablement dignes de notre Illustre Compagnie" (*M*, 73).

31. Frédéric Brandt, *Notice sur la vie de Mr le baron David de Purry, suivie de son testament et d'un extrait de sa correspondance particulière* (Neuchâtel, Switzerland: Wolfrath, 1826), 1, writes: "Mr J. Purry avoit fait de bonnes études." I have been unable to verify this information. The range of Purry's readings (which I will examine in detail in the expanded version of this project) is shown by, among other things, his reference to Isaac Bullart's extensively illustrated in-folio work, *Académie des sciences et des arts, contenant les vies, et les eloges historiques des hommes illustres, qui ont excellé en ces Professions depuis environ quatre siècles parmy diverses nations de l'Europe*, 2 vols. (Brussels: chez F. Foppens, 1682).

32. See, for instance, Carlo Ginzburg, *The Cheese and the Worms: The Cosmos of a Sixteenth-Century Miller* (Baltimore: Johns Hopkins University Press, 1980), 62–65, on how a Friulian miller, from a widely different time, space, and social background, read Genesis.

33. "Car de dire que les hommes ne peuvent pas se resoudre si facilement à quitter leurs liaisons, leurs amis, leurs parens, tout cela ne sont que de niaiseries et des chimères qu'on se met dans l'esprit" (SM, 19).

34. "Du bon goût de leurs melons, du fumet de leur perdrixs, et de tant d'autres choses qui rendent la vie delicieuse" (SM, 19).

35. "Pour moi j'entends par un bon païs, un païs qui abonde non seulement en laict et en miel, mais généralement en toutes les choses capables de flater la volupté et de nous faire vivre delicieusement; un païs de cocagne et de bonne chere, qui est fertile, et qui produict facilement, sans beaucoup de travail et à bon marché, tout ce qui est necessaire à la vie; voilà en peu de paroles et suivant mes petites idées, ce que c'est qu'un bon païs" (M, 22).

36. William Temple, "Upon the Gardens of Epicurus; or, Of Gardening, in the Year 1685," in Five Miscellaneous Essays, ed. Samuel Holt Monk (Ann Arbor: University of Michigan Press, 1963), 12: "And 'tis great pity we do not yet see the history of Chasimir, which Mounsieur Bernier assured me he had translated out of Persian, and intended to publish, and of which he has given such a taste in his excellent memoirs of the Mogul's country." See also Clara Marburg, Sir William Temple: A Seventeenth-Century "Libertin" (New Haven: Yale University Press, 1932).

37. Temple, "Upon the Gardens of Epicurus," 18; Purry quotes from Les oeuvres mêlées de Monsieur le chevalier Temple, 2d ed., 2 vols. (Utrecht, the Netherlands: Chez Antoine Schouten, 1694).

38. See Heeres, Het Aandeel der Nederlanders in de Ontdekking van Australie, 1606–1765, xvi.

39. See Jean-Pierre Purry, A Memorial Presented to His Grace My Lord the Duke of Newcastle (1724; Augusta, Ga.: J. H. Estill, 1880), 1.

40. See Roulet, "Jean-Pierre Purry et ses projets de colonies en Afrique du Sud et en Australie," 55.

41. "Il réalisa la meilleure partie de son bien et courut à Paris, où il spécula avec tant de succès, qu'il possédait un jour dans son portefeuille des effets au porteur pour plus de six cent mille francs. Jean Chambrier, son ami, plus tard ministre de Prusse à Paris, le conjurant de faire comme lui, et de réaliser au moins deux cent mille francs pour les faire parvenir à sa femme et à ses enfants, Purry lui répondit froidement: 'On ne parle ici que de millions, il faut donc aller aux millions, puis nous réaliserons'" (F. A. M. Jeanneret et J.-H. Bonhôte, Biographie neuchâteloise, 2 vols. [Locle, Switzerland: E. Courvoisier, 1863], 2:251). See also Brandt, Notice sur la vie de Mr le baron David de Purry, 1–2.

42. See V. W. Crane, The Southern Frontier, 1670-1732 (1929; Ann Arbor: University of Michigan Press, 1956), 284n8, which refers to B.M. Add. mss. 32,739 (Newcastle Papers, LIV), ff. 39, 41 f. (Purry, letter to Walpole, 6 June 1724, and Walpole, letter to Newcastle, 7 June 1724).

43. See Migliazzo, "A Tarnished Legacy Revisited," 237.

44. See the letters of May 11, 1727, and January 1, 1717, in Recueil de quelques lettres et documents inédits concernant David de Purry et sa famille, 16–17, 13–14.

45. See the French translation of the original patent in the Archives de l'État, Neuchâtel, Switzerland, Archives de la famille de Purry, G. XII.

46. See Jean-Pierre Purry, "Proposals by Mr. Peter Purry of Neufchatel for Encouragement of Such Swiss Protestants as Should Agree to Accompany Him to Carolina, to Settle a New Colony, 1731," in *A Description of the Province of South Carolina* (Washington: Force, 1837), 14–15; see also Jean-Pierre Purry, *Description abrégée de l'etat présent de la Caroline meridionale* (Neuchâtel, Switzerland: Sr. Jacob Boyve at Neufchatel and Sr. Sécretaire Du Bois at St. Sulpy, 1732); and Jean-Pierre Purry, *Description abrégée de l'etat présent de la Caroline meridionale, nouvelle edition, avec des eclaircissemens, les actes des concessions faites à ce sujet à l'Auteur, tant pour luy que pour ceux qui voudront prendre parti avec luy. Et enfin une Instruction qui contient les conditions, sous lesquelles on pourra l'accompagner* (Neuchâtel, Switzerland: Sr. Jacob Boyve at Neufchatel and Sr. Sécretaire Du Bois at St. Sulpy, 1732), 36. *A Description of the Province of South Carolina* is partially republished in Jean-Pierre Purry, *Tracts and Other Papers, Relating Principally to the Origin, Settlement, and Progress of the Colonies in North America, from the Discovery of the Country to the Year 1776*, ed. Peter Force, 2 vols. (Gloucester, UK: Peter Smith, 1963). A much shortened version appears in Jean-Pierre Purry, *Kurtze, iedoch zuverlassige Nachricht von dem gegenwärtigen Zustand und Beschaffenheit des Mittägigen Carolina in America oder West-Indien, welche Landschaft Georgien genennet wird, aufgesetzet in Charlestown oder Carlstadt von vier glaubwürdigen Schweitzern, und aus der Französischen Sprache anietzo verdeutscht. Welchem eine Nachricht von denen so genannte Bilden, welche in derselben Gegend wohnen, beygefüget ist* (Leipzig: Samuel Benjamin Walthern, 1734), 16. For additional bibliographical references, see Jon Butler, *The Huguenots in America: A Refugee People in New World Society* (Cambridge: Harvard University Press, 1983), 217–20.

47. Purry, *Description abrégée de l'etat présent de la Caroline meridionale*, 8, 28.

48. See H. D. K. Leiding, "Purrysburg: A Swiss-French Settlement of South Carolina, on the Savannah River," *Transactions of the Huguenot Society of South Carolina* 39 (1934): 32. This latter text was possibly based on A. H. Hirsch, *Huguenots of Colonial South Carolina* (Durham, N.C.: Duke University Press, 1928).

49. See Max Weber, *The Protestant Ethic and the Spirit of Capitalism*, trans. Talcott Parsons (1930; London: Routledge, 1993); hereafter abbreviated *PE*. For the German, see Max Weber, "Die protestantische Ethik und der 'Geist' des Kapitalismus," *Archiv für Sozialwissenschaft und Sozialpolitik* 21 (1905): 1–54; a revised edition appears in Max Weber, *Gesammelte Aufsätze zur Religionssoziologie* (Tübingen, Germany: Mohr, 1920–21). Parsons translated *innerweltliche Askese* as "worldly asceticism" (see *PE*, 193–94); in his introduction, Anthony Giddens speaks of "'this-worldly asceticism'" (*PE*, xii).

50. See E. Sestan, introduction to *L'etica protestante e lo spirito del capitalismo*, by Max Weber, trans. Piero Burresi (Rome: Leonardo, 1945), xlv.

51. *PE*, 233n68.

52. I owe this suggestion to Alberto Gajano.

53. *PE*, 47.

54. *PE*, 183.

55. In his *The Destruction of Reason* (a much maligned book, in which ideological platitudes and profound passages coexist), György Lukács wrote: "German sociology's stumbling block is the primitive accumulation of capital, and the workers' violent separation from the means of production" (György Lukács, *Die Zerstörung der Vernunft* [Neuwied am Rhein, 1962], 525, my translation). Weber, Germany's foremost sociologist and Lukács's former mentor, was of course the main target of this critical remark. On a more general (and less interesting) level, see Karl Löwith, "Max Weber und Karl Marx," *Archiv für Sozialwissenschaft und Politik* 67 (1932): 53–99, 175–214. The essay was translated and included in *Marx, Weber, Schmitt* (Rome: Laterza, 1994). "C'est probablement Marx qui a exercé sur Weber l'influence la plus profonde et la plus durable," writes E. Fleischmann, "De Weber à Nietzsche," *Archives Européennes de Sociologie* 5 (1964): 194, but without developing the implications of his own remark.

56. Karl Marx, *Capital*, trans. Ben Fowkes, 3 vols. (New York: Vintage, 1977), 1:873–74; hereafter abbreviated as *C*.

57. *PE*, 20. On the mutual exclusiveness between capitalist adventurers and intramundane ascesis, see Max Weber, "Antikritisches Schlusswort zum 'Geist des Kapitalismus,'" *Archiv für Sozialwissenschaft und Sozialpolitik* 31 (1910): 554–99, 596–97, partially republished in Eduard Baumgarten, *Max Weber: Werk und Person* (Tübingen, Germany: Mohr, 1964), 173–91.

58. But he changed his mind on this just before his death, as I will show in the expanded version of this project; see Max Weber, *Economy and Society: An Outline of Interpretive Sociology*, ed. Guenther Roth and Claus Wittich, trans. Ephraim Fischoff et al., 3 vols. (New York: Bedminster Press, 1968), 1:137–38. After having stressed the rationality of capitalist production, Weber remarks: "The fact that the maximum of *formal* rationality in methods of capital accounting is possible only where the workers are subjected to domination by entrepreneurs, is a further specific element of *substantive* irrationality in the modern economic order." The meaning of this remark is clarified by a later passage: "Willingness to work on the part of factory labor has been primarily determined by a combination of the transfer of responsibility for maintenance to the workers personally and the corresponding powerful indirect compulsion to work, as symbolized in the English workhouse system, and it has permanently remained oriented to the compulsory guarantee of the property system. *This is demonstrated by the marked decline in willingness to work at the present time which resulted from the collapse of this coercive power in the [1918] revolution*" (Ibid., 1:153; emphasis added].

59. *C*, 1:925.

60. *C*, 1:917, 918.

61. *C*, 1:882n9.

62. *PE*, 47, 22.

63. See Jacques Revel, "Micro-analyse et construction du social," in *Jeux d'échelles: La Micro-analyse à l'expérience*, ed. Revel (Paris: Gallimard, 1996), 15–36.

64. A single case analyzed in depth will suffice to provide the basis for an extensive comparison; see Marcel Mauss, "Essai sur les variations saisonnières des sociétés eskimo: Étude de morphologie sociale," 1906, in *Sociologie et anthropologie*, 3rd ed. (Paris: PUF, 1966), 389–477.

65. Marcel Proust, *The Guermantes Way*, vol. 3. of *In Search of Lost Time*, ed. D. J. Enright, trans. C. K. Scott Moncrieff and Terence Kilmartin (New York: Chatto and Windus, 1992), 450. "Les niais s'imaginent que les grosses dimensions des phénomènes sociaux sont une excellente occasion de pénétrer plus avant dans l'âme humaine; ils devraient au contraire comprendre que c'est en descendant en profondeur dans une individualité qu'ils auraient chance de comprendre ces phénomènes" (Marcel Proust, *Le côté des Guermantes*, vol. 2 of *À la recherche du temps perdu* [Paris: Gallimard, 1959], 330). The passage, which is on Françoise and the Russo-Japanese war, has been quoted by Francesco Orlando, "Darwin, Freud, l'individuo e il caso," *La Rivista dei Libri* 5 (1995): 21.

Afterword:
Reflections on Exemplary Narratives,
Cases, and Model Organisms

MARY S. MORGAN

**WHAT KIND OF SCIENTIFIC OBJECTS ARE CASES
AND EXEMPLARY NARRATIVES?**

According to the traditional philosophies of history and of science, the particular narratives of history sit at the opposite pole to the lawlike claims of natural science. Yet if we begin our reflections from the humanistic perspective rather than the science perspective, we may come to a rather different view. Exemplary narratives and particular cases marked out by a disciplinary community for wider usage are endemic in the human and social sciences. They provide the objects of those sciences and fulfil roles in those fields comparable to the ones we associate with the set of model organisms of biology. The essays in this volume, spanning from the historical and social sciences into the natural sciences of biology and the earth, suggest not only that the forms of investigation and reasoning of history and science may well be more similar than orthogonal but also that we need an altogether more subtle account of their relations.

The particular cases and exemplary narratives used in the human and social sciences are closely written, organised, and analysed descriptions of events, activities, feelings, and relationships. They offer a marked contrast to the way in which scientific models are most often characterised, namely, as simplified and idealized mathematical representations, either of theories or of relations existing in the natural world. Though of necessity some loss of detail and some idealization takes place in the process of case and narrative construction, idealization and simplification are not the objectives. These cases are emphatically not the ideal types of Max Weber's sociology, as Carlo Ginzburg shows, nor the "bland or bizarre" hypothetical cases that Josiah Ober suggests inhabit political philosophy, nor the stripped down *Homo economicus* of John Stuart Mill's political economy (that, incidentally gave impetus to the mathematical models of modern economics). The aim here, on the contrary, is to capture the full complexity of human and social life in the particular case or in the narrative. Representing this complexity is achieved by description, analysis, revision, and redescription. The case accounts of Belle's fantasy in psychoanalysis, the exemplary narrative of the Trobriand *kula* ritual in anthropology, and the long-lived exemplar of Athenian democracy in political science are *highly concentrated* versions of the real life human and social world they represent, in which complexities of relations and of meanings are maintained, if not accentuated, in the rendering process.

Such cases and exemplary narratives may be usefully compared to another kind of model found in the natural sciences, namely, model organisms in biological sciences. These whole life models are neither simplified nor idealized, but domesticated and highly standardized versions of those creatures found in the wild (though, to the extent they learn to live in the laboratory and cohabit with the scientist, they may not be entirely natural).[1] Cases and exemplary narratives are not standardized, domesticated versions of the human and social world, but they are, in significant ways, both similar to and different from the fruit fly, the lab mouse, and other model organisms.

OBJECTS OF INQUIRY, INQUIRY WITH OBJECTS?

Though cases and exemplary narratives differ in salient ways from model organisms, they share the characteristic feature of being objects to be inquired into and objects to inquire with. Initially, the lab mouse was investigated for its own sake, and then as a comparator for similar behavior or effects in human life, but the mouse and the human did not thereby become conflated. Nevertheless, in these biological, human, and social science realms, it is often difficult to divorce these roles and define the object in separation from its analysis and its wider usage. John Forrester shows that Belle's fantasy constitutes an important object for both analyst and patient. For Robert Stoller, the fantasy was an object to inquire into and to inquire with: it enabled him to analyze Belle. But it was equally an object for the patient to enhance her self-understanding and healing, for it embodied her identity. It constituted an object of inquiry for both of them. It also forms an object for the student coming to an understanding of the field; and, in Forrester's hand, it becomes an object for the reader of this volume to gain insight into how a fantasy or dream serves as an exemplary narrative for both patient and analyst. Multiple identities and multiple roles for the object seem not uncommon in these human and social sciences, just as model organisms serve multiple purposes. Sometimes the distinctions between objects of investigation in their own right and instruments for learning about other things seem scarcely viable. In the recent work on genetics as reported by Marcel Weber, in which the genetic material of fruit flies and humans at the molecular level is found to be sufficiently indistinguishable that cloning can take place, the separate identities of the fruit fly and the human, one as object of inquiry and the other as object to inquire into, seem to have dissolved altogether.

Similarity and difference, and their analysis, provide the leverage for cases, narratives, and model organisms to work as objects of inquiry and investigation in the human and life sciences. Belle's fantasy was not a static object: it changed over the period of analysis, and it was the subtle differences and continuing similarities that formed the notable points of development for both the analyst and the patient: in the course of Belle's analysis, the fantasy became the exemplary narrative for both of them. Athenian democracy holds the status of an exemplary event for Western democracies, but this status does not by itself constitute it as an object of political inquiry, for, as Ober argues, "mere historicism" has no value to the political scientist. Rather, the value of such a case depends on the "testing of intuitions" through the exploration of differences and similarities between Athenian democracy and modern political theory. In contrasting the historically real and imperfect with the normative theory of the universal ideal, this activity of inquiring with the exemplary case may prove illuminating or surprising, and so inform both historical understanding and modern theorizing.[2] A similar process of close matching takes place in my Prisoner's Dilemma case in economics, for the process of defining points of similarity and difference in the iteration between game description and situation description is the way that game theory is applied to understand the world and one of the ways in which game theorizing develops. Rachel Ankeny, writing on biology and medical science, makes broader claims: equating cases with models and portraying case-based reasoning as a method depending on analyzing similarity and difference.

Athenian democracy and the Prisoner's Dilemma constitute an experimental unit for the political scientist and the economist much as do the lab mouse or the worm *C. elegans* for the biologist or medical scientist. This view needs further explication, for the method of locating similarity and contrasting differences in this process of matching such cases and exemplary narratives to the world may not look much like the mode of laboratory experiment on fruit flies. However, this analysis of similarity and difference fits the description that Ankeny gives us of how the wiring diagram of the worm's nervous system is used as a benchmark case against which to calibrate normal and abnormal cases of worm. This process in turn can be related to experiments in the traditionally nonlaboratory sciences such as geology and astronomy. In geology, Naomi Oreskes points to the tradition of mimicking experiments, in which outcomes are examined for similarity and difference to the patterns and evidence left by past geological events. Alistair Crombie writes of astronomy as the first of the scientific fields to make use of models, physical objects constructed to represent what was thought to happen in the unobservable parts of the universe.[3] Success in matching the observable was required before any claims to learn about the rest of the system (to "access the inaccessible," as Oreskes puts it) by manipulating or experimenting with the model could be sustained.

Such physical model work in the sciences has now mostly given way to computer-based simulations of mathematical or statistical models, as we see in the sciences of geology and climate study.[4] Oreskes's essay charts the development from mimetic physical models to such simulations in geology. We first see experiments on small-scale imitations of the earth, made out of similar materials and subject to the kinds of interventions hypothesized to create geological phenomena. Where the effects produced matched (mimicked) certain characteristics found in the world, the experiments were taken to demonstrate possible causes. (In one case at least, the effects offered a reinterpretation of the observations.) What constituted "similar" in the mimetic model itself was all important to claims for the demonstration of causes. Scale was critical, and not just scale in size but scale in characteristics—hard rock at the scale of the earth had to be replaced by wax or pancake batter for the scale of the mimetic model. These experiments might have been labeled simulations had the word been used in that form in the nineteenth century, but the line proves very difficult to draw. When scale was quantified, the experiments became mathematical and numerical, and the subsequent leap into modern simulation technology became inevitable. In geology, as in climate science and economic science, many experiments take place on mathematical, numerical, or computer-based models of the relevant parts of the world, not on the world itself: these are investigations into our specially prepared descriptions of the world just as are those of anthropology, political science, and history. Indeed, the methods of simulation and experimenting with models can be broadly understood as the careful and systematic investigation of similarity and difference as certain variations are made in the model in ways that parallel the variation and matching process for Athenian democracy or the Prisoner's Dilemma game. This matching and iterative process reinforces the double sense here that models are objects to inquire into and to inquire with.

The relationship of the model, case, or exemplary narrative to an associated system in the wider world may also change more fundamentally over time. Susan Sperling shows how the baboon troop was first the subject of natural history for its own sake, and only subsequently became the object or model organism used to study humans' social arrangements, language, and so forth.

Where once anthropologists studied rituals to learn about rituals, ritual came to be understood as constituting one of the natures of society, so that in studying ritual, the anthropologist now studies society. The move from object of inquiry to inquiry with objects is most radically understood in Clifford Geertz's suggestion that ritual has now become the microscope, where once it had been the insect on the slide. This reversal has been most explicitly handled in the science fiction of Douglas Adams's *Hitchhiker's Guide to the Galaxy* in which, at the end of a long saga, it turns out that the earth with its human inhabitants was a model system designed by white mice to run an experiment to answer the fundamental question of the universe.[5]

EXEMPLARITY AND REPRESENTING

Within Forrester's discussion of Stoller's account, Belle's fantasy was exemplary because it held the power to represent her case and changing condition, though it may never become an important, oft-quoted case in the discipline. Yet the process by which a narrative becomes exemplary in the annals of psychoanalysis is perhaps rather similar to the way that a particular organism such as the mouse or fruit fly gains the label of model organism. Once constructed, any subsequent use of that case suggests that it has the power to help understand other cases. As usage grows, the individual case may come to seem exemplary, representing in some critical respects many similar such cases, rather in the same way as Ankeny suggests that the index case of medical science is taken as the norm or pattern against which other like cases must be compared. It becomes a disciplinary object—not just an example but the exemplar consistently referred to because it represents many other like cases (the medical index case) or because it represents some ideal (Athenian democracy) or norm (the worm's wiring diagram).

Many fields have a number of exemplary models or narratives that keep reappearing, and each time they do so, their exemplary status is reasserted and grows stronger. Biology, for example, has a number of exemplary cases in the form of model organisms—though they are not all equally useful for everything biologists want to do. For example, Weber discusses how the lab mouse, which for long had been the typical model organism for human comparison in medical science, proved too big and complex for the purposes of experimental comparative work at the genetic level. The fruit fly, a model organism with an unequaled history of research into genetic variation, provided better experimental resources in this field. Weber shows how well-mapped variations at the genetic level in fruit flies enabled researchers to construct maps of variation at another level, the molecular one, in humans by new techniques of cloning that linked both organisms. But even the fruit fly is too complex for other tasks, as Jane Hubbard recounts in her discussion of how the worm was chosen by Sydney Brenner in the original research to map out the relations between genes and behavior.

Ginzburg's case study of Jean-Pierre Purry may or may not become an exemplary narrative—time will tell—but Ginzburg used him to give two important insights into the characteristics of such a narrative. On the one hand, we have an exact account of what it means to be an exemplary narrative in his description of Purry's use of the Bible (a document that surely must be one of the most widely used exemplary narratives): "Purry thought with the Bible, an experience he shared with innumerable individuals before and after him. The Bible gave him words, arguments, and stories; he projected words, experiences, and events into the Bible." An exemplary narrative is one that we see with in a double way—we both see the world through it and we see our experiences in it: it converts our experiments in life into experiences. At the same time, Ginzburg's own commitment to Purry's narrative lies in his belief that the case is significant because it makes us rethink the similarities and differences in the accounts given by Karl Marx and Max Weber of the development of capitalism. Purry's life may have been randomly found, but it does not prove random with respect to accounts of the world, for it provides us with a lever to consider again those two competing accounts. As a case example that fits neither of those sociohistorical accounts, Purry's life history is a narrative with potential exemplary power. Like the molecular genetics of the little fruit fly in relation to human life, this single man's microhistory has the potential to inform us about something much bigger, namely, the development of capitalism.

The double duty we find in the role of the Bible for Purry is an endemic feature for scientists using such objects. The Prisoner's Dilemma is by no means a case with such particularity and detail as either Ginzburg's life history of Purry or the text we know as the Bible. Yet it operates as an exemplary instance of a general result in economic theorizing about strategic situations and for many different and oft-found situations in the economic world, ones in which the invisible-hand argument, that self-interest will lead to good outcomes for all, fails. One small game situation comes to represent a general point. And, in a comparable, but less personal way than Purry's use of his Bible, economists started off using game theory to understand strategic interactions in the social world, yet increasingly such games have come to be the things they see and experience in the world.

The power of cases and narratives to act as exemplary might also be compared to the way in which models fulfil a number of representing roles in the social, human and biological sciences, and other natural sciences.[6] These representing roles are fulfilled in different ways. A model organism such as the lab mouse may be representative *of* the class of things it is a member of (other mice), while, at the same time, the case findings may also prove representative *for* another class (humans) or wider class (mammals). Ginzburg's use of Purry's life history can be understood as a representative of a class of microhistories, and, as we have seen, it can also be representative for a set of conceptual accounts of the development of capitalists. There is naturally less epistemic power in using the model organism of the worm to learn about humans or other organisms it may be representative for than about other worms it is representative of. As Hubbard, for example, writes about the conservation of molecular function: "The proteins encoded by the DNA of one organism will usually act similarly in chemical reactions in the cells of all organisms"—yet the same chemical reactions may have different consequences in other organisms.

A more common terminology—with different implications—used in some other natural sciences asks us to think of these objects that we inquire into and inquire with as being *representations* of something in the natural world. The kinds of stylized, sometimes highly complex, mathematical models used in climate science, geological simulations, and formalized games in economics are usually understood as representations: models built in a very different medium than the object they represent. These scientific representations *of* the world may be used *for* purposes—such as predicting the next recession or earthquake or rainstorm—but the relationship is not one in which the model is thought of as being representative of one class or for another class of events or objects. The medium of representation found in mathematical models differs so much from the real geological or weather events they are taken to represent that observers coming to the discipline may well have natural doubts about the epistemological usefulness of such model objects. For nonbelievers, perhaps, the status of such mathematical simulations, as for example in geology, must always have been doubtful. Even for believers, the inference power of experiments with such representations is necessarily weaker compared to those from experiments with representative whole-life models.[7]

Different forms of cases, exemplary narratives, and models represent in different ways, and these labels of representations and representatives are useful notions to distinguish the different kinds of representing work that gets done. Yet if we look carefully at the objects revealed in the essays published here, we see that some of them work in the interstices of those two kinds of representing. The Prisoner's Dilemma model may offer just such an interesting crossover case. As a mathematical object for manipulation by the economist, it is a representation of situations of strategic interaction under certain rules. But at the same time, as is evident from my article, it is taken to be representative of a class of game (within a well-defined theory-based taxonomy) and representative for a class of situations found in the economy. Ankeny's wiring diagram of the worm seems to embody other crossover characteristics. As a description derived from a small number of examples to typify the nervous structure of a particular kind of worm, it is understood as representative of the nervous system of that kind of worm. But as a diagram on paper, it is also a representation of the nervous structure of that kind of worm: it is not the whole life organism in the way that the lab mouse is, but a rendering into another medium of one abstracted element of its life.

SCIENCES WITHOUT LAWS AND THE AUTONOMY OF CASES

What does a science without laws look like? A science that must map things, find patterns, and classify cases in order to understand its subject matter seems to be a science without laws. A science in which variation relative to the general behavior is so endemic that only the variations count, seems, for all intents and purposes, a science without laws. A science in which the interaction and amalgamation of the laws is so difficult to pull apart that necessity, causal relations, and generality are lost in the thicket also seems to be, to all intents, a science without laws. Take clouds, for example. Acute observers have no doubt been studying clouds throughout human history. They are an everyday phenomenon, and their appearance in the sky leads people to change their plans, for we understand their import. Scientists have provided a taxonomy of cloud types, have mimicked the formation of clouds in the laboratory, have seeded them in the sky to try to force rain, and have proposed lawlike explanations of their behavior. Yet, as Amy Dahan Dalmedico suggests, integrating the phenomena of clouds into climate science models has proven extremely difficult, if not impossible. What hope then for integrating the size of leaves into such a model, let alone providing a lawlike account of the difference leaf size makes to climate behavior?

Ritual, too, is universal, yet it remains recalcitrant to universal account. There appear to be no laws of ritual, though its role has been theorized in functional and in structuralist terms, and it has been subject to hermeneutic interpretation. Ritual is understood to be the source of a society's identity, so much so that, Geertz suggests, we may recognize individuals' loss of identity as a reflection of our modern societies' lack of ritual. The strong particularity of rituals is marked by the fact that exemplary narratives of ritual are named according to their societies and/or their authors, and not according to a line of theorizing or general explanations. The rituals known as, for example, Geertz's Balinese cockfight, Evans-Prichard's Zande poison oracle, and Douglas's Lele pangolin, remain distinct objects that can be used, reused, and reinterpreted for multiple purposes, just as the various model organisms—mice, fruit flies, worms, and yeast—have been used for many different research purposes. In anthropology, and perhaps in history, these cases remain in some sense "owned" by their original authors: serious analysis and reanalysis of the case—the mining of the detailed and laboriously gathered case materials—remains pretty much restricted to the original author. When there come points of recognition within the disciplinary community, when the case becomes one referred to by others, it becomes an exemplary case within the discipline. The loan of such an exemplary case, for purposes of comparison, is licensed to others, but not its reworking. At this point, the point at which the case is used to give insights into other cases, even for the study of such other cases, situations, or events, it is evident that the case has gained a certain degree of autonomy beyond its original place.[8] It is the autonomy or independent status of such cases which are understood as having wider relevance, yet not easily subsumable under laws, that provides a set of reasonably stable objects within a discipline.

The medical index cases discussed by Ankeny are, by contrast with those of anthropology, owned by the discipline. First a case epitomizing a disease or condition on the basis of a particular instance, the index case gains ground as the generic case against which other new cases will be measured. The index case itself is then revised, rethought, and refashioned so that it gains autonomy, cutting free from the original patient and their particulars as its usage continues. The Prisoner's Dilemma game of economics and the model organisms of biology are also codeveloped and therefore co-owned. Indeed, their status as exemplary objects depends on their wide usage, analysis, investigation, and application by many different workers in the disciplinary field. There is a self-reinforcing quality here too: as cases are developed by others, they take on an exemplary status, but they may also gain the autonomy that betokens uses of the model organism or game in areas well beyond the interests of the original research community. The baboon troop offers a good example of this tendency of very particular cases to become both central to a field's theorizing, and therefore exemplary for that field, and yet gain autonomy to become an object for other disciplines. First featuring as a replacement for the "primitive" man with which more "advanced" human primates were compared, the baboon troop then became the model for evolutionary accounts of human society according to, first, a structuralist-functionalist account, and then a sociobiological account. Taking the baboon troop as a model of social and political behavior allowed for both hierarchical Hobbesian and feminist network interpretations. Baboon troop behavior has been subjected to explanations from anthropology, zoology, evolutionary psychology, sociobiology, and economics, and yet limits to its usage remain, for cultural patterns are not shared among the primates. The extraordinarily flexible lifestyle of such a case is both a sign of and dependent on its autonomy: the baboon troop has become an object to be transported and reused to further understanding in other sciences well beyond its home base.

Sciences dependent on simulations, model organisms, cases, and exemplary narratives indicate that they are, perhaps only for the moment or for certain purposes, sciences without usable laws.[9] But such sciences are not, as we have seen from the essays in this volume, sciences without knowledge, understanding, relevance, or usefulness. Rather, the human, social, and natural scientists of these disciplines investigate particular cases, exemplary narratives, and model organisms in modes that uncover relations, variations, similarities, and differences, and then use these to develop meaningful accounts. As we have seen, they do not just inquire into those objects but use them to inquire with, and thus develop broader and deeper understandings within their fields. Not all models and narratives support these broader investigative uses. Those that gain exemplary status are able to do so because they are taken to say something about a wider set of particular cases or situations than the one from which they grew. This wider relevance indicates how such objects gain the autonomy to function more broadly as instruments of inquiry. In these human, social, and natural sciences, such models, cases, and exemplary narratives seem to provide the stability that some other sciences find in laws.

NOTES

My reflections on the papers in this volume draw in addition on recent work (referenced below) inspired by my participation in the Princeton Workshop in the History of Science titled "Model Systems, Cases and Exemplary Narratives," 1999–2001. My heartfelt thanks go to the organizers of the workshops, Angela Creager, Liz Lunbeck, and Norton Wise for welcoming my participation and inviting me to have the final word in this volume.

1. See Rom Harré, "The Materiality of Instruments in a Metaphysics for Experiments," in *The Philosophy of Scientific Experimentation*, ed. Hans Radder (Pittsburgh: Pittsburgh University Press, 2003), 19–38.

2. In contrast, the recent development of the "analytic narrative" in economic history

tests out historical narratives against economic analysis, but with the aim of extending the historical explanation, not to affect economic theorizing. See Robert H. Bates et al., eds., *Analytic Narratives* (Princeton: Princeton University Press, 1998).

3. Alistair Crombie, *Styles of Scientific Thinking in the European Traditions: The History of Argument and Explanation Especially in the Mathematical and Biomedical Sciences and Arts*, vols. 1–3 (London: Gerald Duckworth, 1994).

4. Two useful resources in the science studies literatures on simulations are Sergio Sismondo and Snait Gissis, eds., "Practices of Modeling and Simulation," special issue, *Science in Context* 12 (1999); and M. Norton Wise, ed., *Growing Explanations: Historical Perspectives on Recent Science* (Durham, N.C.: Duke University Press, 2004).

5. Douglas Adams, *The Hitchhiker's Guide to the Galaxy* (London: Pan Books, 1979).

6. Of particular relevance to these claims are Harré, *Philosophy of Scientific Experimentation*; Mary S. Morgan, "Model Experiments and Models in Experiments," in *Model-Based Reasoning: Scientific Discovery, Technological Innovation, Values*, ed. Lorenzo Magnani and Nancy Nersessian (Dordrecht, the Netherlands: Kluwer, 2002), 41–58; and Mary S. Morgan, "Experiments without Material Intervention: Model Experiments, Virtual Experiments and Virtually Experiments," in Radder, *Philosophy of Scientific Experimentation*, 216–35.

7. See Mary S. Morgan, "Experiments versus Models: New Phenomena, Inference and Surprise," *Journal of Economic Methodology* 12 (2005): 317–29.

8. In Mary S. Morgan and Margaret Morrison, *Models as Mediators: Perspectives on Natural and Social Science* (Cambridge: Cambridge University Press, 1999), chap. 2, we argued that the autonomy of models arose from virtues in their construction. The account here suggests that autonomy is a disciplinary-based and almost social feature of models.

9. These may be sciences with lawlike regularities, but not ones that can be predicted or easily used to explain findings or to apply to cases. This is a particular problem in cases in which scientific activity responds to demands from strong patrons, as occurs in geology, climate science, and medical science. For example, biologists working on the worm provided discoveries about the genetic basis for mutations in reproductive cells, and though such findings may offer insight into the formation of anatomical structures, they do not necessarily provide insights into human reproduction failure. Another image was given recently by Nancy Cartwright in her *The Dappled World: A Study of the Boundaries of Science* (Cambridge: Cambridge University Press, 1999), suggesting that laws only apply to specific places and cases.

RACHEL A. ANKENY is a senior lecturer in the School of History and Politics at the University of Adelaide, Australia. She was formerly director of the Unit for History and Philosophy of Science (HPS) at the University of Sydney from 2000–2004, and senior lecturer in HPS and the Centre for Values, Ethics and the Law in Medicine from 2003–2006. Her published research and ongoing major projects cross several disciplinary areas including the history and philosophy of contemporary life sciences, bioethics, and public policy.

ANGELA N. H. CREAGER is a professor of history at Princeton University, where she specializes in the history of modern biomedical research. She authored *The Life of a Virus: Tobacco Mosaic Virus as an Experimental Model, 1930-1965* (2002) and is currently working on the legacy of the bomb project for postwar biology and medicine through a study of the Atomic Energy Commission's radioisotope program.

AMY DAHAN DALMEDICO is the director of research at the Centre National de la Recherche Scientifique (CNRS) and the joint director at the Centre Alexandre Koyré (EHESS-CNRS). Specializing in the history of the mathematical sciences in their social and political milieus, her works include *Mathématisations: Augustin-Louis Cauchy et l'École Française* (1992) and *J-L. Lions, un mathématicien d'exception entre recherche, industrie et politique* (2005). Currently, she is focusing on applied mathematics, dynamical systems, modeling practices, and climate change.

JOHN FORRESTER is a professor of history and philosophy of the sciences at the University of Cambridge. He is the author of *Language and the Origins of Psychoanalysis* (1980), *The Seductions of Psychoanalysis* (1990), (with Lisa Appignanesi) *Freud's Women* (1992), *Dispatches from the Freud Wars* (1997), and *Truth Games* (1997). With Laura Cameron he is currently completing *Freud in Cambridge*, scheduled for publication in 2007.

CLIFFORD GEERTZ, an anthropologist, was a faculty member of the School of Social Science at the Institute for Advanced Study, Princeton, from 1970 to 2006. He is the author, among other works, of *The Religion of Java* (1960), *The Interpretation of Cultures* (1973), *Local Knowledge* (1983), and *Available Light* (2000).

CARLO GINZBURG for many years taught at the University of Bologna and at the University of California, Los Angeles. He is now a professor of storia delle culture europee at the Scuola Normale Superiore in Pisa. His most recent books include *History, Rhetoric, and Proof* (1999); *No Island Is an Island. Four Glances at English Literature in a World Perspective* (2000); *Wooden Eyes: Nine Reflections on Distance* (2001); and *Il filo e le tracce: Vero falso finto* (2006).

E. JANE ALBERT HUBBARD is on the faculty of the Department of Biology at New York University, where her laboratory uses molecular genetics approaches to understand development of the nematode worm, *C. elegans.* Her work focuses on the developmental genetics of gonadogenesis and germline development, both of which are essential for reproduction. Her laboratory has identified and characterized evolutionarily conserved genes involved in the control of *C. elegans* germ cell proliferation and differentiation.

ELIZABETH LUNBECK is the Nelson Tyrone Jr. Professor of American History at Vanderbilt University. She is the author of *The Psychiatric Persuasion: Knowledge, Gender, and Power in Modern America* (1994), the coauthor, with Bennett Simon, of *Family Romance, Family Secrets: Case Notes from an American Psychoanalysis* (2003), and the coeditor of several collections of essays. She is currently completing a book on narcissism.

MARY S. MORGAN is a professor of the history of economics at the London School of Economics and a professor of history and the philosophy of economics at the University of Amsterdam. She is currently researching both the development of mathematical modeling and the changing nature of observation in economic science. Her published works on the history of economics include *The History of Econometric Ideas* (1990) and two edited volumes, *From Interwar Pluralism to Postwar Neoclassicism* (with Malcolm Rutherford, 1998), and *The Age of Economic Measurement* (with Judy L. Klein, 2001).

JOSIAH OBER is the Constantine Mitsotakis Professor of Political Science and Classics at Stanford University. His books include *Mass and Elite in Democratic Athens* (1989), *Political Dissent in Democratic Athens* (1998), and *Athenian Legacies* (2005). His current research explores the relationship between collective action and the organization of useful knowledge in democracies. Before coming to Stanford in 2006, he taught at Princeton and Montana State Universities.

NAOMI ORESKES is a professor of history and a member of the Program in Science Studies at the University of California, San Diego. Her research focuses on the historical development of scientific knowledge, methods, and practices in the earth and environmental sciences, and on scientific consensus and dissent. She is the author of *The Rejection of Continental Drift: Theory and Method in American Earth Science* (1999), editor of *Plate Tectonics: An Insider's History of the Modern Theory of the Earth* (with Homer Le Grand, 2001), and she is currently completing *Science on a Mission: American Oceanography in the Cold War and Beyond.*

SUSAN SPERLING is a biological anthropologist who has researched and written about primate social behavior and human evolution, women's reproductive development in an evolutionary perspective, and the history of the animal rights movement. She is currently working on a biography of the late evolutionary anthropologist Ashley Montagu. Sperling teaches in the anthropology and interdisciplinary studies programs at

Chabot College (Hayward, California) and in the medical anthropology program at the University of California, San Francisco.

MARCEL WEBER earned a master's degree in molecular biology from the University of Basel, a PhD in philosophy from the University of Constance, and a *Habilitation* in philosophy from the University of Hanover. He is currently a Swiss National Science Foundation professor of the philosophy of science at the University of Basel. His books include *Die Architektur der Synthese: Entstehung und Philosophie der modernen Evolutionstheorie* (1998) and *Philosophy of Experimental Biology* (2005).

M. NORTON WISE is a professor of history at the University of California, Los Angeles, where he also codirects the Center for Society and Genetics. He has been working recently on the commercialization and politicization of science, the gender of automata in nineteenth-century Britain, and landscape gardens in Berlin. He is completing a book titled *Bourgeois Berlin and Laboratory Science*.

Library of Congress Cataloging-in-Publication Data
Science without laws : model systems, cases,
exemplary narratives / edited by Angela N. H.
Creager, Elizabeth Lunbeck, and M. Norton Wise.
p. cm.—(Science and cultural theory)
Includes bibliographical references and index.
ISBN 978–0-8223–4046–1 (cloth : alk. paper)
ISBN 978–0-8223–4068–3 (pbk. : alk. paper)
1. Biological models. 2. Biology—Philosophy.
I. Creager, Angela N. H. II. Lunbeck, Elizabeth.
III. Wise, M. Norton.
QH324.8.S25 2007
570.1—dc22
2007006303